STRATEGIES TO THE PREDICTION, MITIGATION AND MANAGEMENT OF PRODUCT OBSOLESCENCE

WILEY SERIES IN SYSTEMS ENGINEERING AND MANAGEMENT

Andrew P. Sage, Editor

A complete list of the titles in this series appears at the end of this volume.

STRATEGIES TO THE PREDICTION, MITIGATION AND MANAGEMENT OF PRODUCT OBSOLESCENCE

BJOERN BARTELS
ULRICH ERMEL
MICHAEL PECHT
PETER SANDBORN

WILEY

A JOHN WILEY & SONS, INC., PUBLICATION

Published by John Wiley & Sons, Inc., Hoboken, New Jersey
Published simultaneously in Canada

Library of Congress Cataloging-in-Publication Data:

Strategies to the prediction, mitigation and management of product obsolescence / Bjoern Bartels... [et al.].
p. cm.
Includes bibliographical references.
ISBN 978-1-118-14064-2 (hardback)
1. Industrial electronics—Maintenance and repair. 2. Electronic instruments—Design and construction. 3. Product obsolescence. I. Bartels, Bjoern.
TK7881.S905 2012
681'.2—dc23

2011043920

10 9 8 7 6 5 4 3 2 1

Contents

Preface

Engineers and managers must be aware of the life cycles of the parts they incorporate into their systems. Otherwise, they can end up with a product whose parts are not available or a product that cannot perform as intended, cannot be assembled, and cannot be maintained without high life cycle costs. While technological advances continue to meet product development needs, engineering decisions regarding when and how a new part will be used and the associated risks differentiate the winners and losers.

This book will enable manufacturers and supporters of products and systems to manage the obsolescence of the parts that compose their products and systems. This book is intended for engineers and managers, product team members, marketing professionals, business development professionals, and contract negotiators.

This book explains the life cycle of parts and software and presents a process for obsolescence forecasting based on sales data, case studies illustrating forecasting methods, and explanations of reactive, proactive, and strategic obsolescence management strategies.

Chapter 1 describes general definitions and the fundamental issues associated with the occurrence of obsolescence and its management. This chapter builds the foundation for obsolescence management to reduce the risks affecting various products and industries.

Chapter 2 describes the change management methods and controls commonly used by semiconductor manufacturers and the types of changes that they make. Relevant standards and guidelines are introduced and described. Some of the major change management standards development bodies are discussed and examples are given.

Chapter 3 describes the electronic part life cycle from design and introduction to obsolescence. The six stages of an electronic part life cycle are explained and described in terms of attributes such as sales, price, usage, part modification, number of competitors, and profit margin.

Chapter 4 explains several methodologies for forecasting obsolescence. Methodologies based on sales curve forecasting and procurement life analysis are included.

Chapter 5 illustrates the application of the obsolescence forecasting methodology in the form of case studies for different part types. For each of these part types, information on the part type, market trends, procurement life cycle of the part type, and zone of obsolescence are presented.

Chapter 6 discusses software obsolescence. Obsolescence management is not just a hardware problem; it is a hardware and software problem. Hardware changes drive software obsolescence and vice versa.

Chapter 7 explains reactive strategies that can be employed by equipment manufacturers to combat the problem of obsolescence. Reactive obsolescence management is concerned with determining an appropriate, immediate resolution to the problem of components becoming obsolete. This chapter also provides a guide to select an appropriate reactive obsolescence management strategy.

Chapter 8 illustrates strategies to proactively manage obsolescence and track procurement life cycle information on selected parts to prevent obsolescence-driven risks such as production stops and expensive redesigns.

Chapter 9 explains strategic obsolescence management to enable strategic planning, life cycle optimization, and long-term business case development for the support of systems by using obsolescence data, logistics management inputs, technology forecasting, and business trending. This chapter also provides a guide for implementing strategic obsolescence management within an organization.

Chapter 10 describes relevant standards and guidelines for the management of obsolescence. Some of the major change management standards development bodies and organizations are discussed and examples are given.

Finally, an extensive list of references is provided to aid the reader in finding additional information.

Chapter 1

Introduction to Obsolescence Problems

Obsolescence is the status given to a part when it is no longer available from its original manufacturer. The original manufacturer's discontinuance of a part may have many causes, including nonavailability of the materials needed to manufacture the part, decreased demand for the part, duplication of product lines when companies merge, or liability concerns. The problem of obsolescence is most prevalent for electronics technology, wherein the procurement lifetimes for microelectronic parts are often significantly shorter than the manufacturing and support life cycles for the products that use the parts. However, obsolescence extends beyond electronic parts to other items, such as materials, textiles, and mechanical parts. In addition, obsolescence has been shown to appear for software, specifications, standards, processes, and soft resources, such as human skills.

This chapter describes general definitions and the fundamental issues associated with the occurrence of obsolescence and its management in order to build a consistent basis for this topic. Because obsolescence is most prevalent for electronics, this chapter concentrates on the issues associated with obsolescence in relation to electronic parts; however, most of what follows is also applicable for nonelectronic parts as well.

1.1 DEFINITION OF OBSOLESCENCE

The English word *obsolescence* is derived from the Latin term *obsolescere*, which means "to go out of use or fashion." The associated adjective *obsolescent*

Strategies to the Prediction, Mitigation and Management of Product Obsolescence, First Edition. Bjoern Bartels, Ulrich Ermel, Michael Pecht and Peter Sandborn.

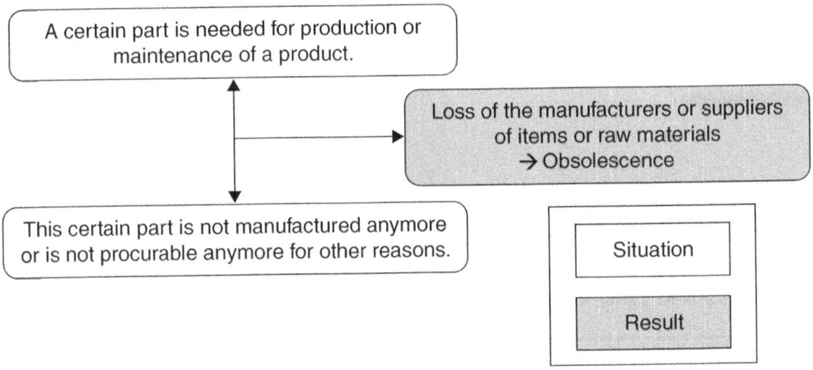

FIGURE 1-1 Appearance of obsolescence.

is derived from the Latin term *obsoletus*, meaning "worn out" (Baer and Wermke, 2000).

Obsolescence, as addressed in this book, refers to materials, parts, devices, software, services, and processes that become non-procurable from their original manufacturer or supplier. As parts become obsolete, users and customers are inevitably faced with a supply shortfall when their demands for the original part cannot be satisfied and no alternate parts are procurable (Atterbury, 2005; Rogokowski, 2007).

Generally, obsolescence is defined as the loss, or impending loss, of the manufacturers or suppliers of items or raw materials, as shown in Figure 1-1 (Tomczykowski, 2001).[1] However, a more realistic working definition of obsolescence is when a part (material or technology) that is needed to manufacture or support a product or system is not available from existing stock or the original manufacturer of the part (material or technology).

There are many possible reasons for obsolescence. Some of the causes of obsolescence include the following:

- Rapid technological development makes a product or part unusable for technical, economical, or legal reasons (Feldmann and Sandborn, 2007)
- The original component manufacturer (OCM) or original equipment manufacturer (OEM) disappears from the market for various reasons (Atterbury, 2005)

[1]This definition of obsolescence is sometimes called "procurement" or "DMSMS-type" obsolescence, where DMSMS stands for Diminishing Manufacturing Sources and Material Shortages. Note: Other definitions of obsolescence that are not relevant to the topic of this book include "sudden" or "inventory" obsolescence, which refers to the obsolescence of an inventory of parts that remain after the demand for the part disappears (Brown et al., 1964). Sudden obsolescence is the opposite of the problem addressed in this book.

- The OCM or OEM is not willing to continue producing a part for economic reasons (usually precipitated by a drop in demand for the part) (Atterbury, 2005)
- Chemical or physical aging processes of parts placed in storage can destroy parts or make it impossible to use existing part inventories in products

Terms such as *obsolescence* and *obsolete* are already used by some companies when they provide a product change notification (PCN) or end-of-life (EOL) notice. In such cases, the part is sometimes still procurable for a limited time; that is, customers may have the opportunity to buy parts one last time and store enough of them to meet their systems' forecasted lifetime requirements. These actions are referred to as life-of-type (LOT) buys, lifetime (last time) buys (LTBs), or bridge buys (see Chapter 7).

1.2 CATEGORIZATION OF OBSOLESCENCE TYPES

The subject of this book is involuntary obsolescence, where neither the customer nor the manufacturer necessarily wants to change the product or the system. Involuntary obsolescence can be categorized as follows (Feldmann and Sandborn, 2007; Rai and Terpenny, 2008):

- *Logistical* Loss of the ability to procure the parts, materials, manufacturing, or software necessary to manufacture and/or support a product.
- *Functional* The product or subsystem still operates as intended and can still be manufactured and supported, but the specific requirements for the product have changed; as a result the product's current function, performance, or reliability (level of qualification) become obsolete. For consumer products, functional obsolescence is the customer's problem; for more complex systems (such as avionics) it is both the manufacturer's and customer's problem. For complex systems, the functional obsolescence of a subsystem is often caused by changes made to other portions of the system.
- *Technological* More technologically advanced components have become available. This may mean that inventory still exists or can be obtained for older parts that are used to manufacture and support the product, but it becomes a technological obsolescence problem when suppliers of older parts no longer support them.
- *Functionality Improvement Dominated Obsolescence (FIDO)* Manufacturers cannot maintain market share unless they evolve their products in order to keep up with competition and customer expectations (manufacturers are forced to change their products by the market). Note that this differs from functional obsolescence in that for commercial products FIDO obsolescence is forced upon the manufacturers and functional obsolescence is forced upon the customers.

1.3 DEFINITION OF OBSOLESCENCE MANAGEMENT

To ensure a constant qualitative performance, an obsolescence management plan should be improved continually. For example, the Plan-Do-Check-Act (PDCA) cycle shown in Figure 1-2 is an appropriate way to satisfy this goal. Developed by Dr. W. Edwards Deming, the PDCA cycle is also called the Deming Cycle or Deming Wheel (Seghezzi, 1996).

To support continuous improvement, obsolescence management organizations must be provided with adequate resources to support necessary activities that are consistent with the organization's business. The company management (for example, the chief executive officer) is responsible for providing these resources and for establishing an obsolescence management plan within the framework of a dependability management system (IEC-62402, 2004).

The management of obsolescence problems is often referred to as "diminishing manufacturing sources and material shortages" (DMSMS) (Saunders, 2006). As addressed in this book, DMSMS specifically refers to the loss of the ability to procure required materials, parts, or technology.

The process for managing obsolescence is illustrated in Figure 1-3 to mitigate or avoid the impact of supply shortfalls for all types of materials, parts, devices, software, services, and processes during the intended life of a product.

Obsolescence management implies life cycle forecasting and other analyses to identify the effects of obsolescence through all stages of the product life cycle. The cost avoidance associated with various management actions must be

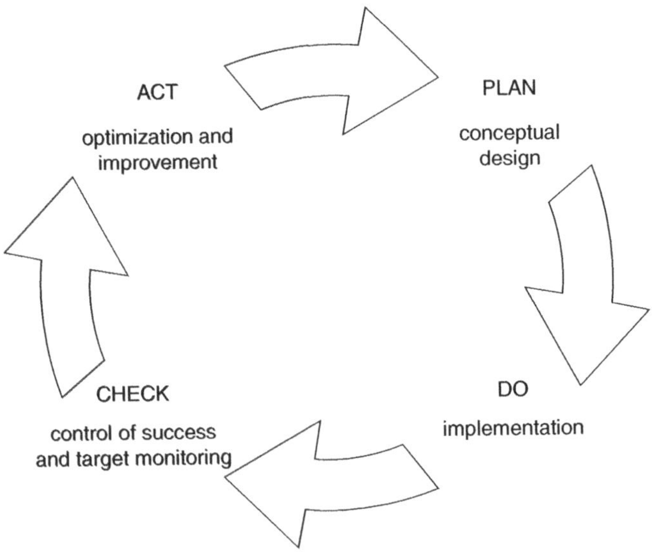

FIGURE 1-2 PDCA cycle.

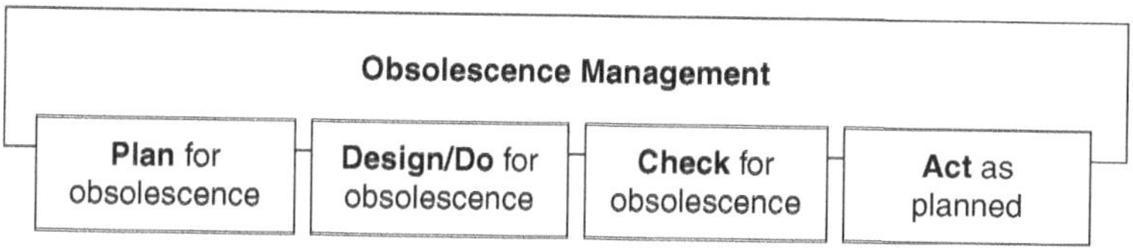

FIGURE 1-3 Process steps for managing obsolescence (adapted from IEC-62402, 2004).

estimated. People must be trained, and resources must be acquired to enable personnel to manage obsolescence. An obsolescence management plan must be developed to ensure adequate selection, timely implementation, and tracking of relevant obsolescence management activities. These activities and other related components and requirements are discussed in the chapters that follow.

1.4 CATEGORIZATION OF OBSOLESCENCE MANAGEMENT APPROACHES

DMSMS require addressing the problem of obsolescence on three different management levels: reactive, proactive, and strategic, as shown in Figure 1-4.

Reactive management (see Chapter 7) is concerned with determining an appropriate, immediate resolution to the problem of components becoming obsolete, executing the resolution process, and documenting/tracking the actions taken. Common reactive DMSMS management approaches include, among others, lifetime buy, bridge buy, component replacement, buying from aftermarket sources, uprating, emulation, and salvage (Sandborn, 2008).

Proactive management (see Chapter 8) is implemented for critical components that have a risk of going obsolete, lack sufficient available quantity after obsolescence, and will be problematic to manage if or when they become obsolete. These critical components are identified and managed prior to their actual obsolescence event. Bill of material (BOM) management regarding obsolete or soon to be obsolete components is an important part of the design and manufacture of any product. Proactive management requires the ability to forecast obsolescence risk for components. It also requires there be a process for articulating, reviewing, and updating the system-level DMSMS status (Sandborn, 2008).

Strategic management (see Chapter 9) of DMSMS means using DMSMS data, logistics management inputs, technology forecasting, and business trending to enable strategic planning, life cycle optimization, and long-term business case development for the support of systems. The most common approach to DMSMS strategic management is design refresh planning, determining the set of refreshes (and associated reactive management between refreshes) that maximizes future cost avoidance (Sandborn, 2008).

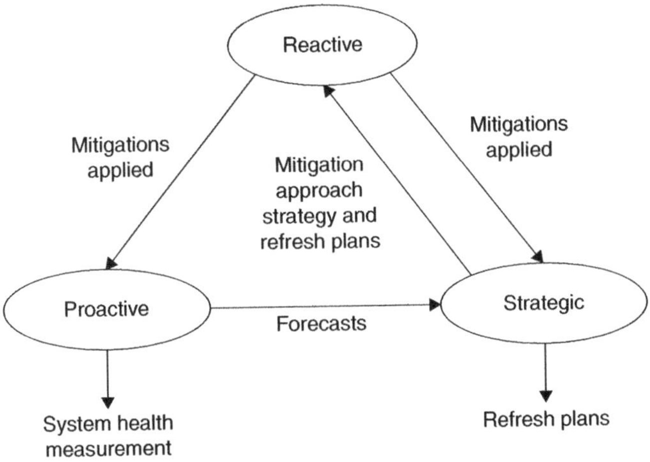

FIGURE 1-4 Three obsolescence management DMSMS categories and the resulting outputs (adapted from Sandborn, 2008).

1.5 HISTORICAL PERSPECTIVE ON OBSOLESCENCE

Although the origins of electronic part obsolescence are often associated with the advent of acquisition reform in the U.S. Department of Defense in the mid-1990s, concerns about general technology obsolescence as it relates to procuring technology can be traced to much earlier times.

It is evident that the concepts associated with procurement obsolescence were noticed in the context of technology as early as the 1970s. In *The Railway Game* (Lukasiewicz, 1976), Lukasiewicz points out that the market environment in which the railway industry operates restricts them to, in many cases, only one supplier, thus creating a plethora of low-volume supply chain problems that include obsolescence issues.

Although the basic concepts of technology procurement obsolescence have existed since 1970 and probably earlier, the first known mention of the problem specifically related to electronic parts was in 1978 (Smith, 1983) and was associated with the transition from vacuum tubes to solid-state electronics.

References to the acronym DMSMS first appeared in the early 1980s when the U.S. Department of Defense began sponsoring electronic part obsolescence workshops and conferences. The usage of the acronym DMSMS is also seen on the cover of the proceedings from the 1983 DMSMS workshop sponsored by the Defense Electronics Supply Center, shown in Figure 1-5.

The first known component obsolescence management guide was prepared for the P-3 Orion, by ARINC in 1984 (Kuehn, 1984).

The commercialization of obsolescence forecasting for electronic parts began at Zeus Components, Inc., and was used to analyze customer parts lists for sourcing support in early 1986. Hughes Aircraft and Westinghouse offered

FIGURE 1-5 Cover of the proceedings from the 1983 DMSMS workshop (courtesy of Walter Tomczykowski, ARINC).

to pay for the service in late 1986. TACTech (Transition Analysis Component Technology) separated from Zeus in early 1987 and became the first commercial provider of obsolescence forecasting for parts (Baca, 2010).

The real shock wave that put DMSMS on everyone's radar screens occurred when Motorola and Intel terminated their military semiconductor businesses in the early 1990s, a move that impacted virtually every U.S. military program (Baca, 2010). This was followed by the Perry Directive (Perry, 1994) in 1994. The Perry Directive states in part:

> We are going to rely on performance standards . . . instead of relying on milspecs to tell our contractors how to build something . . . There will still, of course, be situations where we will need to spell out how we want things in detail. In those cases, we still will not rely on milspecs but rather on industrial specifications [i.e., non-government standards] . . . In those situations where there are no acceptable industrial specifications, or for some reason they are not effective, then the use of milspecs will be authorized as a last resort, but it will require a special waiver.

The Perry Directive does not mandate the use of commercial components; however, in the wake of the Perry Directive, developers of military systems (and systems that relied on the same supply chain as military products), increasingly moved toward commercial off-the-shelf (COTS) parts, thus accelerating obsolescence issues.

Since the late 1990s, many electronic database tools that include obsolescence status and obsolescence forecasting have appeared, as well as other tools for inventory and demand consolidation and strategic refresh planning. These tools will be discussed in the chapters that follow.

1.6 OCCURRENCE OF OBSOLESCENCE

In order to develop an effective plan to combat part or component obsolescence, understanding the nature of the problem is critical. It is essential to understand how obsolescence can occur and the types of obsolescence that exist.

1.6.1 Technological Evolution

A new generation of technology effectively makes its predecessor obsolete. An example of this would be faster microprocessors making slower ones obsolete. Typically, the new generation technology has improved performance and functionality, often at a lower cost than its predecessors.

1.6.2 Technological Revolutions

In a technological revolution, a new technology supersedes (displaces) its predecessor. An example of this is the fiber distributed data interface (FDDI) that is becoming obsolete as the market moves toward adopting fiber channel as the communications technology of choice.

More common examples are the CD-ROM, which has greater storage capacity and speed than the floppy disk, DVD/Blu-Ray discs that have better quality and more multimedia functions than VHS, and the telephone, which enabled audio transmission instead of the coded electrical signals of a telegraph (ComputerInfoWeb, 2010).

1.6.3 Market Forces

Obsolescence due to market forces occurs when the demand for a component or technology falls, and the manufacturer considers it uneconomical to continue production. This is an increasing problem, as low-volume markets no longer have the purchasing power necessary to persuade manufacturers to continue production. Part manufacturers and distributors may not be willing to manufacture or stock parts that have a small market. The cost of managing the distribution of low-volume parts while providing affordable prices is a challenge; hence, the few distributors that do provide low-volume parts charge high fees.

1.6.4 Environmental Policies and Restrictions

Obsolescence can be caused by directives, rules, and other legislation imposed by governments. For example, EC-Directives are regulations of the European Community for all member states to reach specific goals associated with the usage and waste of specific materials.

For example, the following directives have been implemented in recent years:

- The directive on Waste Electrical and Electronic Equipment (WEEE) from 2003 to reduce the electronic scrap going into landfills by increased recovery, reusage, and recycling (Directive 2002/96/EC, 2003)
- The directive from 2003 on the Restriction on Hazardous Substances (RoHS) to ban specific substances in products sold in the EU that could end up in the waste stream (Directive 2002/95/EC, 2003)
- The directive from 2006 on the Registration, Evaluation, Authorization and Restriction of Chemicals (REACH) to regulate chemicals used in products (Directive 1907/2006/EC, 2006)

To illustrate how these directives affect obsolescence, consider the RoHS directive. Through the RoHS, the usage of lead (Pb) is limited to 0.1 percent by weight for products sold in the EU. Consequently, lead-free solder (for example, SnAgCu) has replaced tin-lead solder (ZVEI, 2008).

RoHS applies to the majority of electronic products. Current exemptions from RoHS include medical devices, monitoring and control instruments, and military and aerospace equipment. The reason for these exemptions is that the long-term effects and reliability of lead-free solder have not been determined.

Because of the RoHS directive, many tin-lead solder finish electronic products have been discontinued (gone obsolete). However, the repair and maintenance of products that were manufactured before the RoHS directive requires tin-lead solder finished electronic parts (Brewin, 2005). The current exemptions from RoHS are largely a moot point because the exempted product sectors (due to their low volume) must use the same supply chains as the non-exempt product sectors.

1.6.5 Allocation

Allocation obsolescence is caused by long product lead time, resulting in temporary obsolescence usually categorized as a short-term supply chain disruption. For example, during the worldwide recession in 2008–2009, many manufacturers reduced production and inventory in order to stabilize their businesses. As customers for parts recover and the demand for parts grows, temporary unavailability of parts can result. In addition, in some cases it appears that chip manufacturers may be delaying capital expenditure while enjoying the higher prices (Allocation Components, 2010).

Beginning in 2010, the reluctance to recommission production lines in response to growing demand led to significant increases in lead times and prices for various parts and materials. An example of 2010 lead times for specific electronic parts is shown in Table 1-1; the impact on prices of raw materials is shown in Table 1-2.

TABLE 1-1 Lead Time Prognosis Overview (March 2010)

Product Group	Lead Time (weeks)	Lead Time Prognosis
Power and Filter (DC/DC, AC/DC, etc.)	Up to 14 weeks	Increasing
Batteries – Primary (Lithium ion, etc.)	Average of 10 weeks	Constant
Batteries – Rechargeable (Lithium ion, etc.)	Average of 12 weeks	Constant
Interconnect (Cables, connectors, etc.)	Up to 12 weeks	Increasing
Electromechanical (Switches, fans, etc.)	Up to 20 weeks	Increasing
Passive (Capacitors, resistors, etc.)	Greater than 30 weeks in some cases	Increasing

(adapted from Avnet, 2010)

TABLE 1-2 Cost of Raw Material Prognosis Overview (April 2010)

Product Group	Advance in Prices within 3 Months (April 2010)	Price Prognosis
Raw Iron	Up 21%	Increasing
Steel Scrap	Up 38%	Increasing
Aluminum Cast Material	Up an average of 13%	Increasing
Nonferrous Metal	Up an average of 18%	Increasing
Lead Alloys	Up 9%	Increasing
Synthetic Materials	Up an average of 13.5%	Increasing

(adapted from Pleyma, 2010)

Allocation, in general, is a double-edged sword. On the one side, it allows manufacturers and suppliers to charge higher prices for their products; on the other side it causes short-term supply chain disruptions that need to be managed.

A further example of allocation issues occurred in mid-2010 when China decreased its exports of rare earth elements. China, with a market share of 93 percent, is nearly the only supplier of rare earth elements in the world. Rare earth metals are used in many electronic components (such as capacitors), and as supplies decreased, long lead times and increasing prices were unavoidable (Zuehlke, 2010).

In addition, natural disasters such as the earthquake that struck northern Japan in March 2011 can cause allocation obsolescence of parts and components. The earthquake and subsequent tidal waves (tsunami) affected electronic component manufacturers' employees, power supplies, and infrastructure and manufacturing facilities, making it impossible to operate as usual. As a result, several electronic component manufacturers had to announce temporary unavailability, longer lead times, and shortages of their parts (Allocation Components, 2011).

1.6.6 Planned Obsolescence

Planned obsolescence refers to an assortment of techniques used to artificially limit the durability of manufactured goods in order to stimulate repetitive consumption (Slade, 2007). In 1954, Brooks Stevens, an American industrial designer, popularized the phrase "planned obsolescence." Stevens's definition of planned obsolescence was, "Instilling in the buyer the desire to own something a little newer, a little better, a little sooner than is necessary" (Milwaukee Art Museum, 2010).

The origins of the phrase "planned obsolescence" go back at least as far as 1932, when Bernard London wrote his leaflet, "Ending the Depression through Planned Obsolescence." He blamed the Great Depression on consumers who used their old products, such as cars, radios, and clothing, much longer than statisticians had expected (Adbusters, 2010; APT News, 2010).

Planned obsolescence, also referred to as built-in obsolescence, is a method of stimulating consumer demand by designing products that wear out or become out-of-date after limited use. Manufacturers increase profits by forcing the customer to buy the next generation of the product after a fixed (planned) useful or functional product life cycle (ComputerInfoWeb, 2010). If the manufacturer has a monopoly, or at least an oligopoly, planned obsolescence or built-in obsolescence may be part of their business strategy (Orbach, 2004).

The majority of examples of planned obsolescence can be found in commercial products. In 2003, consumers expected to use their electronic systems for a maximum of two years before purchasing a replacement or upgraded product. Examples of systems that benefit from planned obsolescence include cell phones, PCs, printers, digital cameras, DVD players, LCDs, gaming systems, mp3 players, and many more (Slade, 2007).

The real problem with planned obsolescence appears when commercial off-the-shelf (COTS) parts designed for use in commercial systems with short procurement life cycles have to be used in systems with much longer product life cycles.

1.7 PRODUCT SECTORS AFFECTED BY OBSOLESCENCE PROBLEMS

Increasing globalization and technological progress make markets and production in different countries dependent on one another and rapidly shorten the procurement life cycles of components and products. In the past several decades, technology has advanced swiftly, causing components to have shorter procurement life spans. Driven by the consumer product sector, newer and better components are being introduced frequently, rendering older components obsolete (Sandborn et al., 2007). As a consequence, the risk of components

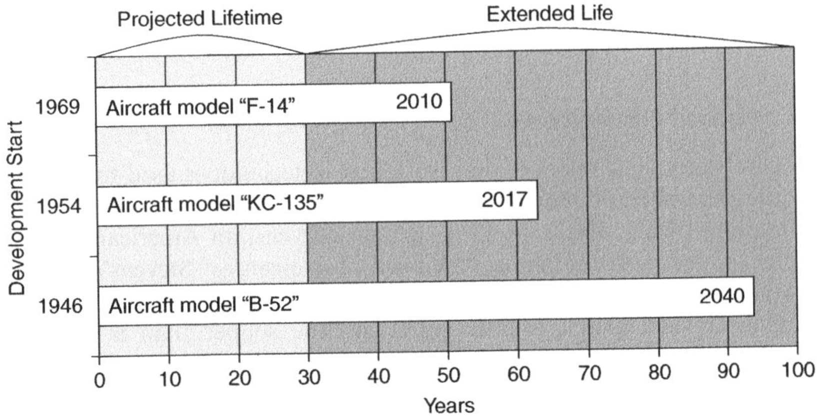

FIGURE 1-6 Extended life of military weapons systems (adopted from Livingston, 2000).

becoming obsolete exists in nearly all product sectors. However, some specific product sectors are affected more than others.

The complexity of the problem is demonstrated in Figure 1-6. This figure shows different military weapons systems that were each designed for a projected lifetime of 30 years. However, many systems for military and defense are being used far longer than originally planned. For example, the B-52 aircraft is projected to operate for more than 94 years, and many weapons systems are expected to have a life span of more than 40 years (Livingston, 2000).

Note that the length of time from the start of design to the beginning of production is increasing. This means that many technologies originally designed into systems are obsolete even before production starts (Hitt and Schmidt, 1998).

The extended life of products and the increasing time period from the start of design to the beginning of production are making it more difficult to supply original spare parts for the whole life span of these products.

Since an increasing number of obsolescence events within the whole product lifetime need to be handled, expenditures on obsolescence management are increasing as well, as indicated by the following examples (McDermott et al., 1999):

- US$81,000,000 was allocated by the U.S. Air Force for the F-22 program to purchase obsolete or soon-to-be out of production parts and to redesign assemblies to accept COTS parts.
- US$600,000 was spent by an avionics manufacturer for commercial airlines to replace an obsolete Intel chip.
- US$500,000,000 was spent to redesign an obsolete radar system for the F-16 program.

- US$264,000 was spent on a life of type (LOT) buy as a resolution for one obsolete logic device for the KC-130F/R program in fiscal year 1997.
- US$250,000 was stated by the deputy under secretary of defense for logistics (DUSD [L]) to be the average cost to redesign a circuit card to eliminate obsolete components.
- US$26,000 to US$2,000,000 is the range reported by the Electronic Industries Alliance (EIA) Manufacturing Operations and Technology Committee as the cost range for redesign of a circuit card.

The longer the product life, the more instances of obsolescence will occur. The product sectors of military and aerospace industries, medical technology, automotive industries, telecommunication industries, and nuclear energy industries are the most affected by obsolescence.

1.8 PARTS AFFECTED BY OBSOLESCENCE PROBLEMS

Obsolescence events are projected to occur more often in the future due to the accelerating pace of innovations. In 1965, Gordon Moore, cofounder of Intel Corporation, noticed that the number of transistors that could be placed on an integrated circuit was doubling about every two years. Furthermore, he predicted that the trend would continue for at least ten years. This forecast (Figure 1-7) is now known as Moore's Law, and advances in integrated circuits still follow it today (Intel, 2010b).

The frequent occurrence of obsolescence in electronics is due to their short procurement life cycles and because the effects on supportability and readiness are generally more immediate and apparent for electronic components. For nonelectronic components, obsolescence problems have generally been slower to develop, and drastic shifts in technology are not as common (Howard, 2002).

In summary, all types of product groups are affected by obsolescence. However, nonelectronic components typically remain supportable for decades, whereas electronic components may become obsolete in a matter of a few years or even months.

1.8.1 Electronic Part Obsolescence

Electronic part obsolescence is generally a result of the rapid growth of the electronics industry. As a result, many of the electronic parts in products have a procurement life cycle that is significantly shorter than the product life cycle of the system they support.

Some examples of the discrepancy between the life cycles of electronic parts and the product lives are shown in Figure 1-8.

The impact of obsolescence can be seen in Figure 1-9, which shows the total number of product discontinuance notices (notices from the original

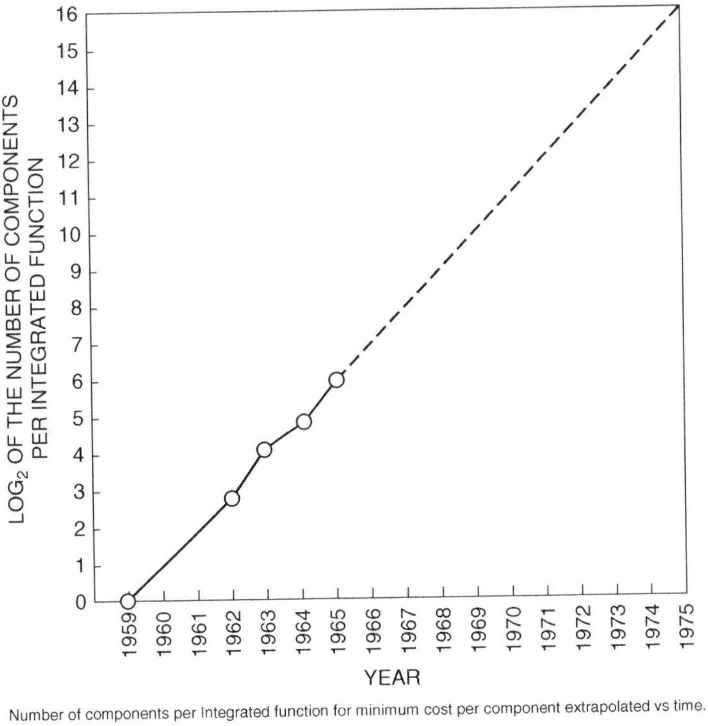

FIGURE 1-7 Graph of Moore's Law created in 1965 (Intel, 2010b).

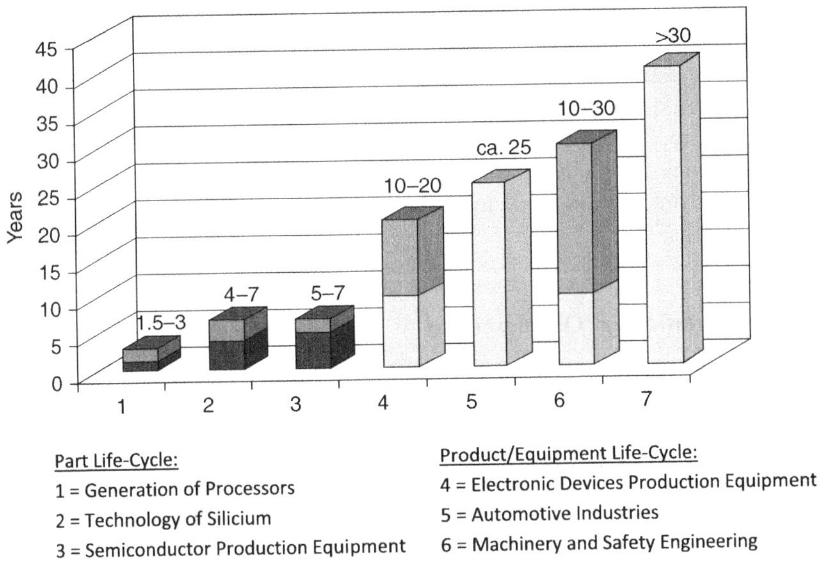

FIGURE 1-8 Discrepancy between part life cycle and product life (adapted from HTV, 2009).

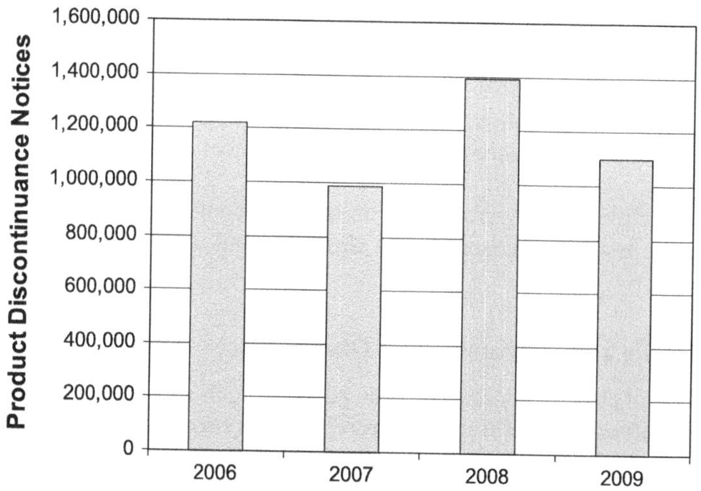

FIGURE 1-9 Numbers of product discontinuance notices for electronic parts (Sandborn et al., 2010).

manufacturer that manufacturing of the part will be terminated) in 2006–2009 from SiliconExpert Technologies, Inc. As of June 14, 2010, SiliconExpert Technologies' parts database consisted of 157,184,671 unique parts (approximately 121.6 million of which are not obsolete), spanning 337 product lines from 11,054 manufacturers. Part count includes all derivations of part numbers based on part family name and generic codes as assigned by their manufacturers. The 1.1 million electronic part discontinuances in 2009 represent approximately 0.9 percent of the electronic parts available in the market (Sandborn et al., 2010).

1.8.2 Software Obsolescence

Software does not wear out, and the cost of generating more copies of software is negligible (IEC-62402, 2004). However, software obsolescence is a significant problem, as the following statement from Bill Gates, founder of the Microsoft Corporation, indicates:

> The only big companies that succeed will be those that obsolete their own products before someone else does. (APT News, 2010)

Software obsolescence is generally due to one of three main causes (Sandborn, 2007):

- Hardware, requirements, or other software changes to the system make the functionality of the software obsolete (functional obsolescence).
- The sales and support for software terminate when the original supplier no longer sells the software as new; when the inability to expand or renew

licensing agreements occurs; or when the software maintenance terminates because the original supplier or third parties no longer support the software (technological obsolescence).

- Digital media obsolescence, formatting, or degradation limits or terminates access to software (logistical obsolescence).

The principles that govern the management of software and hardware obsolescence issues are generally not the same and will be considered later in this book.

1.8.3 Textile and Mechanical Part Obsolescence

Technological change in nonelectronic parts is much slower than for electronic parts and software. However, a comprehensive obsolescence approach has to also contain information on textile and mechanical problems, including the future provisioning of sole-sourced devices. Today, nonelectronic components are also beginning to impact cost and operations through life support issues (Smith, 2000).

Typical symptoms of nonelectronic obsolescence include the following (Howard, 2002):

- For economic reasons, all qualified sources no-bid a component.
- The materials or the manufacturing process become obsolete as suppliers develop stronger, lighter, and more damage-resistant materials.
- Suppliers phase out older materials to meet new production or environmental regulations that have made the use of specific materials illegal.
- The technical data, drawings, or the specifications for procurement are incorrect or incomplete.
- The OEM or OCM goes out of business.
- Specialized tooling, such as casting molds, forging dies, holding fixtures, and sheet metal patterns, required to manufacture the component is unavailable or cannot be refurbished.
- Specialized test equipment is unavailable.

Nonelectronic obsolescence problems for textile and mechanical parts will continue to mount as the systems they support start aging. Inventory within the military, with its product life cycles of over 40 years, is especially affected by this problem (Howard, 2002).

As they apply to software issues, the principles for managing electronic part obsolescence and nonelectronic part obsolescence issues are basically the same. Therefore, the management of textile and mechanical part obsolescence will not be considered further in this book. Management tactics, processes, methods, and procedures referring to electronic parts are comprehensively valid and are also applicable for textiles and mechanical parts.

Chapter 2

Part Change and Discontinuation Management

Change is a natural and inevitable aspect of part manufacturing as companies respond to changing market conditions and technological advancements. The development of new technologies and improved manufacturing processes, constantly changing business forces, and the emergence of new environmental regulations all necessitate changes that a manufacturer may have to make to remain competitive. How a manufacturer manages change can have a large impact on economic success and customer satisfaction. If changes are not implemented in a controlled manner, changes that adversely affect part reliability are more likely to be inadvertently made, damaging a manufacturer's reputation and increasing liability risks. If changes are made frequently, or if insufficient notice or reason is provided for changes, customers could react negatively. Effective change-notification requires part manufacturers to communicate with their customers frequently and openly, such that a bond of understanding can develop. In addition to the careful crafting of communications and management of business relationships, judgment calls are also often made in change-control. The true effects of changes are often unknown, and the distinction between major and minor changes is often fuzzy, despite the presence of industry standards. Often what is considered a minor change to a majority of customers and in the eyes of an industry standard could be a critical change to others using the part in specific applications. Change-control is therefore not only a science, but also an art.

Strategies to the Prediction, Mitigation and Management of Product Obsolescence, First Edition. Bjoern Bartels, Ulrich Ermel, Michael Pecht and Peter Sandborn.

For equipment/product manufacturers, change-control has become increasingly complicated. As captive parts suppliers are divested and supply-chains are becoming increasingly more complex, the amount of control that equipment manufacturers have over the change process has diminished. An increasing number of companies are purchasing parts through distributors, who have no industry standards to guide and promote uniformity in their change-notification processes and who often have varying levels of service from one product/program to another and from one customer to another. The number of paths for the flow of change-notification information through the supply-chain has grown. In today's supply-chain, it is increasingly important for equipment manufacturers to take an active role in the change tracking process. Equipment manufacturers must establish contractual agreements with the manufacturers and distributors from whom they purchase parts in order to ensure that they receive the change-notifications that they require.

2.1 THE CHANGE PROCESS

In most companies, the change process starts with the submission of a proposal to a change-control board (Phillips, 1987), sometimes called an engineering-control board. This board is usually composed of representatives from all major divisions within a company, including marketing, manufacturing, product engineering, and reliability engineering. Any division within the company can propose a change to the board.

Upon receipt of the change request, the board first classifies the change as either major or minor. This classification involves deciding whether the form, fit, or function of the part, as defined by company policies, would be affected by the change. The scope of parts affected, severity of the change, risks involved, and any applicable contractual agreements are also considered. After classification, the board then assesses the associated risks and benefits. Part characterization and reliability stress-testing results are reviewed. If the board determines that the benefits outweigh the risks, the change is approved. However, a change is generally approved only when a convincing business reason for the change exists. If the change is a major change, a notification is sent out to customers. For minor changes, the change is usually implemented without customer notification. For customers that manufacture safety-critical systems, notification for any type of change may be required.

2.2 CHANGE-CONTROL POLICIES OF MAJOR PART MANUFACTURERS

In an effort to evaluate the uniformity of change-control policies in industry, a survey of the top semiconductor manufacturers was performed. It was found that specific change-control board policies vary widely from one company to the next. Many have policies detailing the amount of testing that needs to be done to propose a change to the board, and many have policies on how quickly the changes are phased into production. For example, ON Semiconductor uses a simple chronological system. First, an initial product change notification is issued. After this, the final product change notification confirms the change's implementation (ON Semiconductor, 2010).

Samsung has a three-tiered classification system (Samsung, 2002). The most drastic changes, such as changes in production location, are classified as Class A and require full requalification data. More moderate Class B changes, such as changes in testing procedures, require only "semi-qualification" data. Finally, less significant Class C changes, such as changes to packaging used for shipment or marking on components, require only a simple data review.

National Semiconductor goes even further, with a four-tiered classification system (National Semiconductor, 2002b). Level 1 changes are any changes to a process, material, method, or part design that do not represent a fundamental change to the process technique, material, or part functionality. These changes require no formal qualification or customer notification. Level 2 changes are changes that have only a small chance of impacting part performance and require qualification testing but generally no customer notification. Level 3 changes represent a fundamental change and require a formal qualification, as well as customer notification per contractual agreements. Level 4 changes consist of large-scale part or process transfers that require significant resources and coordination. These changes necessitate a complete requalification, as well as customer notification. National Semiconductor also has three levels of change-control boards, one each to handle changes of Levels 2–4, which incorporate increasingly senior employees on the board for each.

As a final example, the change-control process for IBM Microelectronics is illustrated in Figure 2-1. The IBM Microelectronics change-control board is called a technical review board (TRB), and it ensures that all changes at IBM Microelectronics are made in a controlled manner.

2.3 CHANGE-NOTIFICATION POLICIES OF MAJOR COMPANIES

The change-notification policies of semiconductor manufacturers vary widely, depending on the individual manufacturer, the company division or manufacturing location, and the customer for the parts.

In an effort to evaluate the uniformity of change-notification policies in industry, a survey of the top semiconductor manufacturers, distributors, and contract manufacturers was conducted. The findings of that survey are summarized in the following sections.

2.3.1 Differences by Manufacturer

Most major manufacturers provide change-notifications in a manner compliant with the EIA/JEDEC specifications (see Section 2.4.1.1), despite the common

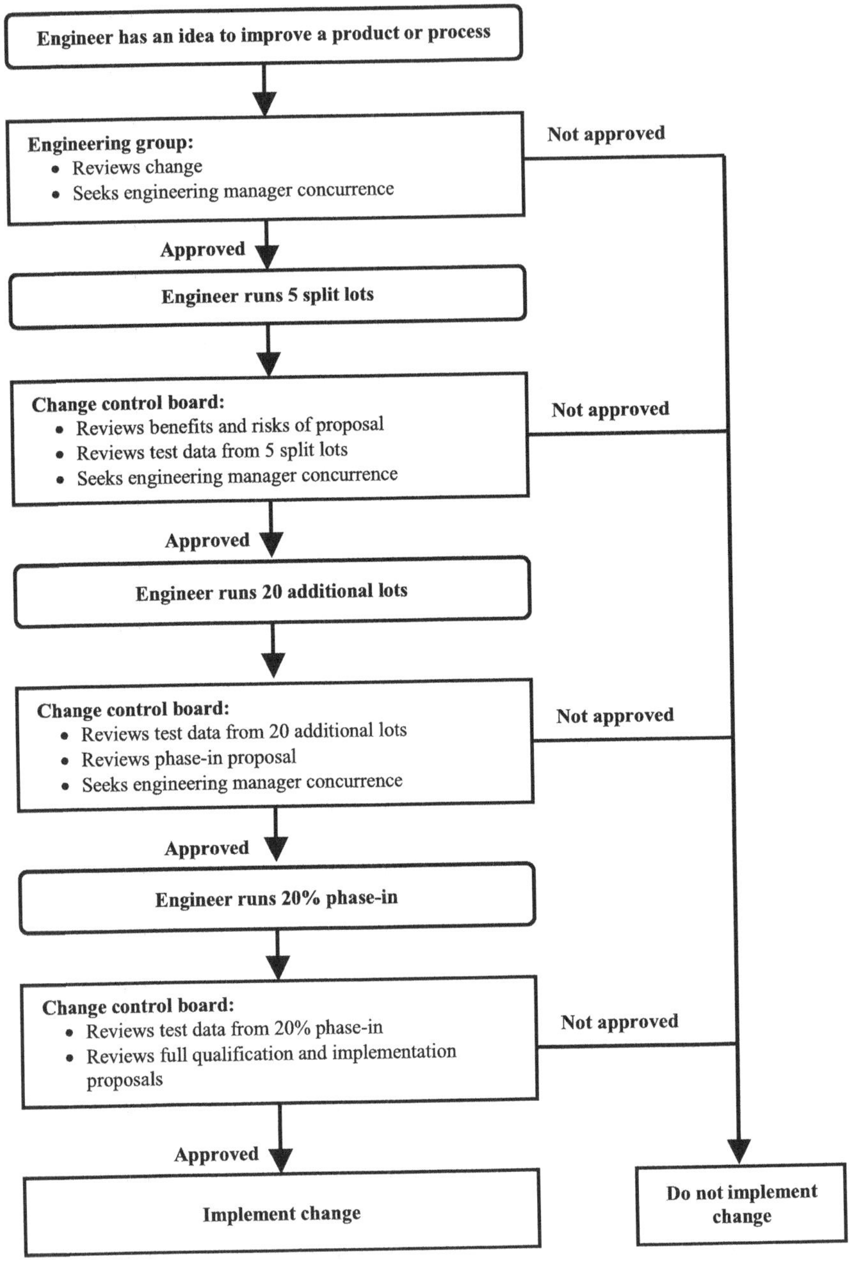

FIGURE 2-1 Change-control process of IBM Microelectronics (adapted from Pennington, 1999).

practice of putting disclaimers on data sheets, which say that the manufacturer is free to make changes or discontinue parts without notice. Changes that could affect reliability or performance are usually sent out 60–90 days in advance, while minor changes, such as marking changes, are usually sent out 30–90 days in advance. Part-discontinuance notices are generally sent out at least six months in advance, although companies with more specialized parts often provide more advance warning. For example, Lattice Semiconductor provides one-year advance notice for last-time buys for its programmable logic devices, per the EIA/JEDEC specification, for parts that are likely to be single-sourced. For extra assurance that customers do not miss a change, some manufacturers, including Fujitsu and Xilinx, also modify part number suffixes after all changes. If, for some reason, a customer missed a change-notification, their shipping dock would likely notice the change when it started receiving parts under an unrecognized part number.

However, there are manufacturers that do not follow the EIA/JEDEC specifications. For example, Samsung Semiconductor sent out a discontinuance notification for a Flash NOR memory family only three weeks in advance (Samsung, 2011).

Analysis of smaller manufacturers resulted in far more variance. Smaller manufacturers are more likely than their larger counterparts to not meet advance notification requirements, not have their policies documented, and not have their quality system audited by external accreditation agencies. In a study performed by CALCE EPSC in the 2000–2001 period, 21 smaller component manufacturers used by a particular equipment manufacturer were audited, many of which were passive component manufacturers. Of these, only 9 were found to have documented and audited change-notification policies. Evaluation of change-notification policies of suppliers is a critical element of an equipment manufacturer's part selection and management program, especially when smaller manufacturers are being considered (Syrus et al., 2001a, 2001b).

2.3.2 Differences by Division or Manufacturing Location

A common company name does not necessarily guarantee that two parts will have the same change-notification policy. Different company divisions or manufacturing locations can have different quality and policy manuals and therefore different policies. Vishay, a large passive part manufacturer, is an example. While Vishay Sprague in Sanford, Maine, has a documented policy in its quality manual to notify customers of part or process changes that affect customer requirements, as of 2001, Vishay Israel in Holon, Israel, was reported to have no such policy (Syrus et al., 2001a). Any time a production location change is made, equipment manufacturers should verify that the policies applied to the parts have not changed.

2.3.3 Differences by Customer Type

Large customers generally have more influence with part manufacturers and therefore can demand more strict and customized requirements from manufacturers than can customers buying smaller quantities of parts. The different industries in which customers operate also result in different change-notification requirements, as different product life cycle requirements exist. In addition, customers that manufacture products in which there is a high degree of liability in the event of a failure, such as automotive or avionics parts, tend to have stricter policies on the handling of change-notifications. For some applications, military customers have even stricter requirements, due to military specifications and procedures that must be followed. Some applications still have requirements for military-grade parts, testing, and change-notification. However, military electronics are increasingly using commercial grade parts. For example, for certain products, Texas Instruments offers enhanced process change notification and obsolescence management. These products are often designed for the military, space, or aviation industry and are referred to as HighRel parts. Texas Instruments refers to these parts as TI-enhanced products. The majority of these parts conform to AQEC (Aerospace Qualified Electronic Components) standards (GEIA, 2005). These standards give basic principles on quality and reliability issues for aviation.

2.3.4 Differences by Geographical Location

Change-notification policies also differ depending upon the country in which the customer is located. Dealing with customers of different cultures requires manufacturers to tailor their policies to match the culture of the country in which they are selling parts (Sullivan et al., 2001). Examples of these differences can be seen when one compares current practices in the United States to those in Japan. As Japanese companies generally prefer more advanced planning than their American counterparts, change-notifications are generally provided with more advance notice to Japanese customers. Where necessary, part manufacturers stock excess inventory in advance of changes to allow longer periods to elapse before Japanese customers receive changed parts. The Japanese also place much more emphasis on personal contact and relationships in business than do Americans. Common practice in Japan therefore calls for part change notices (PCNs) to be translated into Japanese, printed on paper, and hand-delivered by agents of the part manufacturer to the customer. Practices in Western Europe are similar to those in the United States. However, practices vary more and are generally less formal in Eastern Europe. In countries such as Greece and Turkey, business practices are guided more by negotiated understandings than by written contractual agreements. Less emphasis is placed on formal written documentation and rigid change-notification policies, and distributors use more discretion in deciding whether to pass on change-notification information. In addition to cultural influence, these practices are also a product of the legal systems in these countries. Regulations requiring business documentation are generally minimal, and only enough documentation to obtain ISO certification is usually produced. This lack of written notification can complicate the change-notification process, as the change-notifications are often passed on to customers orally. It may require a number of communications for the details of a change to be understood by an end customer.

2.3.5 Distributors

The National Electronic Distributors Association (NEDA) of the United States estimates that currently 35 percent of all North American electronic component sales are made through distributors, and this number is rapidly increasing (NEDA, 2002). Many equipment manufacturers who purchase parts only from distributors are dependent on them for getting part change and discontinuation notices (PCNs).

There are no industry standards guiding the change-notification policies for distributors. Although most distributors carefully track part discontinuance notifications, much less care is taken with part change notifications. Distributors usually have a central office that collects PCNs and passes them on to sales offices. However, it is often up to the individual sales offices to decide to which customers the notifications need to be sent, and practices from one sales office to another can vary widely, even within the same distributor. Policies also vary based on product line and any agreements that a distributor has with a particular part manufacturer. There are also clear service levels within the distributor's customer population, as customers who order large volumes of parts or pay higher prices in exchange for more services get better treatment in the part change and discontinuance notification process. It is, therefore, important for a company to determine what level of service they expect with their sales representatives at distributors, especially if they do not track these changes themselves through an additional notification path. Like the policies for part manufacturers, the policies for distributors also vary from one country to another, reflecting local culture and business customs (Sullivan et al., 2001).

2.3.6 Contract Manufacturers

The largest contract manufacturers in 2009 were Foxconn and Flextronics (Venture Outsource, 2011). Some contract manufacturers are now providing supply-chain management services to their customers and are taking a more active role in part selection and qualification than they have in the past.

Policies of contract manufacturers vary depending on agreements with individual customers and the level of involvement of the contract manufacturer in part selection and management. Larger contract manufacturers with active part selection and management organizations usually have a centralized PCN tracking and notification system. These notification policies are similar to those recommended for semiconductor manufacturers in EIA/JEDEC standards JESD46 and JESD48 (Flextronics, 2011).

2.4 CHANGE-NOTIFICATION

Several key standards and authorities guide the notification processes for semiconductor part changes. However, change notifications can propagate a number of different ways through the supply chain, and, despite the presence of industrywide standards, exact policies and procedures related to change-notification vary.

2.4.1 Industry Standard Process Change-Notification

There are several standards that have been developed to bring uniformity to change-control and notification processes. Industry standards have been developed by organizations such as EIA/JEDEC, which form the basis for most change-control and notification procedures. The U.S. military also has specifications on change-control for military electronics. However, they are increasingly adopting industry standards as the number of military specification parts being manufactured continues to decrease.

Manufacturers of electronic parts customarily decide the content, recipients, and frequency of product change notices. Some equipment manufacturers complain that they are not kept fully informed of changes that affect the functionality of final products. At the other extreme, other equipment manufacturers complain that they are burdened with too many—often trivial—change notices that do not add value to their products and processes. Users and manufacturers need to come to a mutually beneficial, technologically feasible, and contractually enforceable agreement regarding the issuance and delivery of PCNs. Some of the possible means by which a supplier can improve the effectiveness of the notification process are discussed next.

Equipment manufacturers should include detailed conditions about PCNs within all sales contracts or long-term agreements they enter into with part manufacturers (or distributors). The agreement should include (but not be limited to) the following information:

1. Mutually acceptable definitions of terminology for commonly used technical terms (form, fit, function, and so on).
2. Special distribution and content requirements to fit into the equipment manufacturers' part selection and management program.
3. Conditions for issuance of notice (types of changes that prompt a notice).
4. Standard notification timing.
5. Separate conditions for advance notices for parts used in critical applications.

6. A built-in methodology for updating the use profiles of parts in different products and applications.
7. Conditions concerning the duration of and responsibility for record retention.

A well-established practice based on professional trust and ethics can help industry prepare better for handling obsolescence. In particular, it can improve the traceability of parts, help in identifying failure mechanisms, and ease the process of design modification due to changes in components. Some examples of changes that should require the issuance of PCNs are listed in Table 2-1.

2.4.1.1 Electronic Industries Alliance The Joint Electron Device Engineering Council (JEDEC), a member organization of the Electronic Industries Alliance (EIA), has published Standard JESD46-A "Guidelines for User Notification of Product/Process Changes by Semiconductor Suppliers" (JESD46A, 1994). This standard establishes guidelines for customer notification of part and process changes by semiconductor suppliers.

JESD46-A defines part changes, establishes change-notification policy requirements, describes the content of written notifications, and suggests how these procedures can be customized for individual customers. Under this standard, documents called part (or process) change notifications form the basis of the change-notification process.

TABLE 2-1 Examples of Changes That Might Require User Notification

Design	Major Design Change	
Wafer fabrication	Wafer fabrication site	Wafer diameter
	Process flow	Diffusion dopant
	Gate oxide material	Gate oxide thickness
	Dielectric material	Polysilicon dopant type
	Metallization material	Metallization thickness
	Passivation material	Passivation thickness
	Die coating material	Die coating thickness
Assembly	Assembly site	Lead frame base material
	Plating material	Wirebond method
	Mold compound material	Sealing material
	Die attach material	Marking method
	Marking appearance	Plating technique
Test	Elimination of test stops	
Electrical specifications	Changes in DC specifications	
	Changes in AC specifications	
Mechanical specifications	Changes in case outline	Loosening tolerances
Packing, shipping, and labeling environment	Carrier (reel, tray) dimensions	Drypack requirements
	Maximum storage temperature	

(Adapted from JESD46A, 1994)

TABLE 2-2 Examples of Major Changes

Type of Change	Examples of Changes Commonly Made
Fabrication	Manufacturing site, process flow, materials, wafer diameter, mask
Assembly	Assembly site, materials, marking, package style
Test	Test elimination
Electrical specification	AC or DC data sheet parameters
Mechanical specification	Case outline, package tolerances
Packing, shipping, labeling	Carrier tape dimensions, maximum storage temperature, drypack requirements

(Adapted from JESD46A, 1994)

EIA/JEDEC classifies changes as either major or minor. Major changes affect form, fit, function, quality, or reliability and necessitate the notification of all customers. Examples of changes that EIA/JEDEC considers major are provided in Table 2-2. Minor changes are defined as those that do not affect these factors. Notification policies for minor changes depend on individual company policies and requirements specified by customer contracts, but usually no notification is provided. Examples of minor changes are the replacement of an old transfer molding machine with an equivalently functioning new one, resequencing of tests performed on parts following manufacture, and recalibration of equipment.

Compliance to Standard JESD46-A requires a manufacturer to have a documented and controlled change-notification procedure. This procedure should detail the methods of identifying customer notification requirements, notification timing, contents of change-notification documents, methods of documenting delivery of notifications to customers, record retention periods, and methods of obtaining customer approval of proposed changes. The classifications of types of changes should also be defined in a controlled document. The specification recommends 60 days' advance notice be provided to customers before they start receiving a changed product, and that records on changes be maintained on file for a minimum of five years.

Some companies send out two sequential notifications—first an advance notification and then a standard notification. Advance notifications are not required unless a special agreement with a customer exists, but they are useful in soliciting input from key customers when a change is being considered. Standard notifications follow up the advance notifications and include more specific information about the change. Recommended content for notifications includes part identification (part numbers, product lines, and so on), description of change, PCN number, reason for change, implementation date, effects on quality and reliability, methods for distinguishing old parts from new parts, qualification data, and a person to contact with any questions. Customer part numbers may also be included to customize the notification for major customers.

Major customers may establish special guidelines with suppliers governing the change-notification process. Where special agreements exist, they override the standard procedures of the supplier. Customer profiles (also called user profiles) can be generated for each customer to aid in managing the change-notification process. Elements to be considered in customer profile preparation include the part numbers purchased, commodity types, approval requirements and contacts, advance notification requirements, and special contractual agreements. Such agreements built into contracts ensure that a company gets as much control over changes as possible.

2.4.1.2 U.S. Military Military electronic equipment manufacturers buying parts to specifications MIL-PRF-38535 (1989), MIL-PRF-38534D (1999), and MIL-PRF-19500M (1999) have required stricter change-notification procedures. They are based on the Qualified Manufacturers List (QML) and Government Industry Data Exchange Program (GIDEP). QML dictates change-notification and test requirements, whereas GIDEP serves as a recommended repository for change-notification.

QML establishes the general performance requirements for semiconductor parts and the quality and reliability requirements that must be met. The manufacturer notifies the qualifying activity prior to the implementation of any change to a product or quality assurance program that may affect performance, quality, reliability, or interchangeability. The manufacturer submits engineering data, quality conformance inspection data, or a suggested test plan. The manufacturer also explains the qualifying process that ensures that reliability and quality will not be affected and that the product will continue to meet the specification requirements.

GIDEP is a program run by the U.S. government and aids in change-notification. Participants in GIDEP are provided electronic access to engineering data, failure experience data, metrology data, part information data, reliability and maintainability data, and PCNs. To aid in obsolescence management, diminishing manufacturing sources and material shortages (DMSMS) notices of part discontinuances and information on alternate sources and aftermarket suppliers are provided.

2.5 CHANGE-NOTIFICATION PATHS

The routes by which PCNs reach users also vary widely. Low-volume manufacturers will usually buy through distribution and so may not receive the notices directly. They must rely on their distributor to pass them on. Users who have not bought any parts for a long time may also be dropped from the notification list, even though they may still have future requirements to support equipment in service. Some part manufacturers operate a notification service by e-mail, and any user can sign up to receive them. However, these notices are often just part of the general change-notification system operated by the part manufacturer,

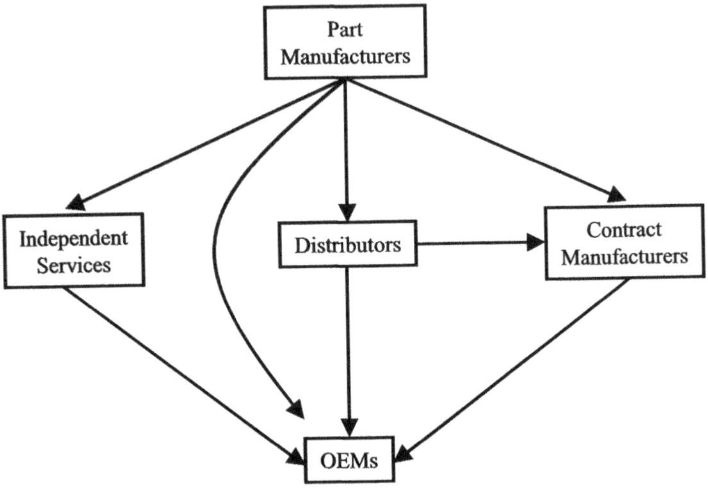

FIGURE 2-2 Electronic part change notification paths.

and the number of e-mails can be overwhelming, leading to the risk of missing an important end-of-life (EOL) notice.

There are four major paths for the propagation of a PCN through the supply-chain to the equipment manufacturer. These paths are illustrated in Figure 2-2. The choice of paths depends on the manufacturer and customer involved. Some manufacturers send out notifications along multiple paths for maximum assurance that the message will reach its destination.

2.5.1 Direct to Equipment Manufacturers

The first path for PCNs is direct notification of equipment manufacturers by the part manufacturer making the change. For large customers who buy parts directly, this is the predominant notification method. This method may also be used for customers who have special notification requirements. Most large manufacturers also maintain PCN websites that can be directly searched by the public. For example, Texas Instruments has a website that lists part change-notifications for each part type, categorized by date of issue and change type, with an abstract summarizing each change (Texas Instruments, 2001a).

2.5.2 Via Distributors

Since most customers buy their parts through distributors and not directly from manufacturers, this method is the most commonly used change-notification path. Manufacturers send all of their PCNs to their distributors. The distributors are then responsible for reviewing sales records and making sure all appropriate customers are notified of part changes. This procedure might take more time than direct propagation from the manufacturer. Moreover, the majority of distributors do not register sample purchases as regular orders in their sales records. Therefore, change- or discontinuance-notifications might not be forwarded to the equipment manufacturer. It is possible for a new product to be qualified with parts that, unbeknownst to the OEM, change prior to the start of manufacturing. Another aspect is the number of different formats for the PCNs. All distributors use their own standard change- or discontinuance-notification format. In addition, using multiple distributors (often done to optimize the purchasing conditions) leads to multiple notifications for the same part.

2.5.3 Via Contract Manufacturers

Equipment manufacturers that outsource manufacturing functions may receive some of their change-notifications from contract manufacturers. In this case, the contract manufacturer receives a change-notification from a part manufacturer or distributor, identifies the products that use the part, and passes that information on to the appropriate equipment manufacturers. Equipment manufacturers may also receive notifications of assembly process changes on their products made directly by the contact manufacturers through the same line of communication. For the equipment manufacturer, this is a very comfortable solution, because all issues discussed in Section 2.5.2 are the responsibility of the contract manufacturer.

2.5.4 Via Independent Services

Independent change-notification services are also available, such as PCNalert and other commercial services. These services collect change information from a large number of part manufacturers and redistribute it to equipment manufacturers. Engineers can search PCNs, view PCN history for various part numbers, and set up an e-mail alert based on specified filters. Information can be filtered against a bill of materials (BOM) or searched based on the manufacturer or type of change (such as shipping/labeling, design change, and discontinuance). GIDEP is considered to be an independent notification service as well, one that is targeted to military customers.

2.6 EXAMPLES OF COMMON CHANGES

Semiconductor technology, packaging materials, and business environments are constantly evolving. New regulations, such as lead-free and halogen-free product initiatives, also bring about changes in electronic parts. Keeping track of current trends is important for equipment manufacturers, both for planning technology insertion and for part selection and management, in order to minimize the overall life cycle costs associated with their products and remain competitive in the marketplace. Explanations of many common part changes are provided in the following sections.

2.6.1 Fabrication Changes

A fabrication change involves moving the manufacture of semiconductor wafers from one location to another. Fabrication changes are often made to accommodate changing production volumes or to minimize costs (Texas Instruments, 2000a). To keep operating margins high, fabrication plants need to be kept as close to full capacity as possible. Fabrication plants may also be purchased or sold due to changing business needs, which can result in part changes or discontinuances. When Texas Instruments closed its fabrication plant in Hatogaya, Japan, in late 2000, for example, many parts sourced from the facility were discontinued (Texas Instruments, 2000b). Introduction of new product lines can also lead to fabrication changes, as production locations of existing product lines are rearranged to make room for them. Acquisitions also continue to be a popular strategy for corporate growth (Ahuja and Katila, 2001), leading to realignment of manufacturing capabilities and operations. Fabrication changes occur particularly frequently for electronic part manufacturers who do not own their own fabrication facilities and must utilize pure-play foundries.

2.6.2 Die Revisions

A die is the piece of semiconductor material (usually silicon) within an IC within which electrical circuitry is fabricated. A large number of circuits are made on a piece of semiconductor material called a wafer and then cut up (a process called dicing) into the individual die. The most frequent die changes are mask changes and die shrinks.

Mask changes are defined as a layout modification on the die and are often the result of process changes. A mask change is often made to add new functionality to a part or improve performance, such as adding improved electrostatic discharge (ESD) protection circuitry (Texas Instruments, 1999b). A mask change may also be made to allow for processes that increase throughput rates (ON Semiconductor, 1999; 2000a), improve manufacturability (Texas Instruments, 2000c; 2000d), or improve testability. Additionally, mask changes may be needed because of fabrication changes, as a part must change to meet the capabilities and process flows of the fabrication plant in which it is built (Xilinx, 2000).

Die shrinks are defined as the scaling down of a mask to produce a smaller die size. Die shrinks lead to higher manufacturer profits, as more die can fit on a single wafer. This often leads to improved electrical performance due to the shorter interconnect lengths. Microprocessor manufacturers such as Intel perform very frequent die shrinks. Having just finished converting all production to 45nm technology, Intel moved rapidly to the next die shrink in 2010, converting virtually all of its production to 32nm technology (Intel, 2010f). Die shrinks are often performed simultaneously with mask changes (Texas Instruments, 1999b; Xilinx, 2000).

2.6.3 Changes to Assembly/Test Locations

Electronic part manufacturers are increasingly outsourcing the assembly and testing portions of their manufacturing processes. These companies want to concentrate their investments in areas where they have competitive advantages. Outsourcing also makes it possible to gain access to specialized knowledge and innovations outside the boundaries of the company. Furthermore, using external suppliers enables the company to increase its ability to react quickly to innovations or changes in market demands (Nielsen, 1999). As high-volume manufacturing of commodity products has traditionally been a low-margin business, outsourcing these functions can also lead to a more attractive balance sheet for a company (Kline, 2001). Part manufacturing flows are often moved from one location to another to meet changing levels of demand and minimize costs (Texas Instruments, 2001b).

Semiconductor manufacturers are also always seeking out new regions of the world to set up manufacturing operations to minimize labor costs. In earlier days, American companies built factories in areas of the United States with lower wages and manufacturing costs, such as Arizona and Texas. For many decades after that, East Asian nations such as Taiwan, Malaysia, and South Korea were preferred. As costs in these regions are increasing, new locations are now being sought, including China (ON Semiconductor, 2000b), the Philippines (Texas Instruments, 1999c), and Thailand (Texas Instruments, 2001b). Some manufacturers are now building manufacturing facilities in Eastern Europe and the former Soviet Union (Carbone, 2001). For example, ON Semiconductor moved some manufacturing operations to the Czech Republic (ON Semiconductor, 2000a). Asian manufacturers are also setting up manufacturing operations in the Balkans, mainly to manufacture products destined for European Union countries (Farish, 1995). At the same time, however, manufacturing is also starting to move back to developed countries such as the United States for some types of products. As the amount of automation used in newer types of semiconductor products is increasing, the portion of manufacturing costs resulting from direct labor is decreasing. If the manufacture of a product is largely automated, the cost savings that could be saved in labor by manufacturing overseas is often more than offset by the additional transportation and logistics costs and decreases in overall company operational efficiency that are incurred (Kline, 2001).

2.6.4 Changes to Assembly Materials

As a part moves from one assembly house to another, changes must be made to the part to adjust it to the process flow and capabilities of the new assembly house. For example, Texas Instruments changed the die attach and lead frame in their TO-220 package to match the process flow of PSI Technologies Inc., one of their assembler/testers (Texas Instruments, 1999c). Changes may also be necessary in order to improve the reliability of parts if problems are encountered in the field. For example, Texas Instruments increased the solder ball size on their MicroStar Junior™ Ball Grid Array packages to improve solder joint and board-level reliability (Texas Instruments, 2001c).

New resin materials, as well as new additives, are constantly being developed to lower cost, reduce moisture absorption, improve processability, and improve electrical and mechanical properties. Lower stress and faster flowing encapsulants are constantly being developed to improve part reliability and reduce costs and are a necessity for some newer, smaller package styles such as ball grid arrays (BGAs) and chip scale packages (CSPs). Special encapsulants are even being developed that offer increased compatibility with copper die metallization, which is starting to replace aluminum in some types of semiconductor parts.

Many changes in assembly materials will also be occurring in the coming years as a result of the introduction of halogen-free technologies. The conversion to lead-free products is a good example of material changes (Palmer, 2002; Robins, 2002). Encapsulants need to be developed that can tolerate the higher processing temperatures required to melt lead-free solders for attachment to circuit boards. Texas Instruments has also qualified a new Sumitomo encapsulant material in preparation for the conversion to lead-free manufacturing (Texas Instruments, 2000e). Another example of material change is the conversion from aluminum-based bonding to copper bonding, as demonstrated in several products from Texas Instruments in 2010 (Texas Instruments, 2011b).

2.6.5 Packing, Marking, and Shipping Changes

Changes to the markings on a part and the packing materials used for shipping are frequently made. Although such changes do not impact part performance, they can wreak havoc on an equipment manufacturer's receiving and manufacturing departments if proper advance accommodations are not made. Examples include changes between laser, ink, or paper part labeling (Intel, 2000a), changes to make labeling more legible or to improve traceability (Intel, 2000b), and changes to the number of parts packed in each reel or tray within a box (Texas Instruments, 2000f).

Packing changes are also made for environmental reasons. Waste reduction and increased recycling are additional green efforts currently popular in the industry. For example, in order to promote the reuse of boxes used for shipping parts to customers, Texas Instruments stopped placing its logo on its boxes in 1999 (Texas Instruments, 1999d). Moisture sensitivity labeling is also being increasingly applied to packing materials as the use of lead-free technologies has grown (Texas Instruments, 1999e).

Chapter 3

Introduction to Electronic Part Product Life Cycles

The electronics industry is one of the most dynamic sectors in the world economy. In the United States, this industry has grown at a rate three times that of the overall economy since the 1990s. The rapid growth of the electronics industry, the massive competition among part manufacturers, the high investments in production lines, and bankruptcies attibutable to the economic crisis of 2008–09 have together cumulatively spurred a dramatic change in the manufacturing of the products and systems we buy. New parts are constantly being introduced with increased speed, reduced feature size and supply voltage, and leading-edge interconnection and packaging technologies. As a result, new products and systems that use these new parts are exploiting these capabilities both to provide product differentiation and to increase demand and sales. Even as the worldwide per capita income decreased by 2 percent in 2009 in comparison to 2008, the number of smart phones sold grew by 13.2 percent, the number of shipped LCD TVs by 42 percent, and the number of netbooks shipped by 100 percent (Markt & Technik, 2010a).

As described in Chapter 1, for product sectors that require a long use life, the electronic parts that comprise a product have a procurement life cycle that is significantly shorter than the support life of the product they are in. It is this product life cycle mismatch between the parts and the product that is the root of the obsolescence problem.

As an example, consider the engine control module used in a car. The BMW Z3 1.9i, manufactured in 2002, has an engine control module that is still

Strategies to the Prediction, Mitigation and Management of Product Obsolescence, First Edition. Bjoern Bartels, Ulrich Ermel, Michael Pecht and Peter Sandborn.

available even though production of the control module ended in 2002. However, the Windows NT–based hardware platform for programming these control modules (required for the central locking system) is no longer available at most authorized workshops. Many similar examples can be found in other industries, such as defense and energy electronics, avionics, industrial controls, medicine, and railroads. A private passenger automobile has a typical life of ten years, and thus replacement parts and controllers are required to be available for this period. The engine controller manufacturer depends on the same suppliers as high-volume industries such as the personal computer market, where the product life cycle is typically two to three years and the underlying technology of the parts used has a similar procurement life cycle. Part technology moves on during this period, preparing for the next high-volume product. Moreover, part manufacturers will consolidate their manufacturing facilities to service the new higher-volume products. This process leaves the original part without a manufacturing base, thus rendering it obsolete. Of course, there are some parts that remain available for long periods of time, but due to the high costs of requalification, even a single obsolete part can cause expensive problems.

This product life cycle mismatch problem requires that engineers be cognizant of which parts will be available and which will be obsolete when the product is manufactured and supported. This awareness is especially important for engineers in the automotive, avionics, and defense industries, as well as other industries where there is a long-term support obligation. These systems may encounter obsolescence problems before being fielded and invariably experience obsolescence problems during field life (Bumbalough, 1999). Parts that are substituted into the product after the original design may require assembly processes to be changed in order to use the new parts. If newer parts with changed performance characteristics are used as substitutes, the designers must anticipate the changes to ensure that circuit timing, noise margins, and electromagnetic interference (EMI) non-conformances do not suddenly create severe problems at the product level. If the product requires a long application life, then an open architecture and proactive design techniques, or a part obsolescence strategy (lifetime buy, aftermarket purchases, or part substitution) may be required.

Part selection needs more critical attention, because the root cause of obsolescence, at any product level, can be the obsolescence of one single part. Figure 3-1 shows the possible impact of obsolescence on the design process of a sonar system. For instance, Figure 3-1 shows that 70 percent of the commercial off-the-shelf (COTS) parts in this system are already obsolete before the first system is installed (Singh and Sandborn, 2006).

3.1 PRODUCT LIFE CYCLE STAGES

Most electronic parts pass through several product life cycle stages corresponding to changes in part sales. Figure 3-2 (Pecht and Das, 2000) is a representative

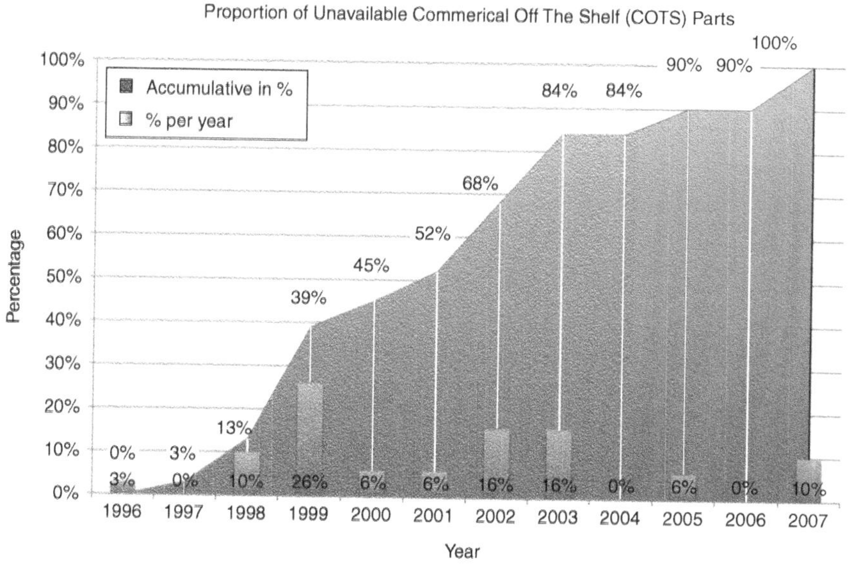

FIGURE 3-1 Percent of commercial off-the-shelf (COTS) parts that are out of production (un-procurable) during the first ten years of the surface ship sonar system's life cycle. (Courtesy of NAVSURFWARCENDIV Crane) (Singh and Sandborn, 2006)

product life cycle curve of units shipped per unit of time, which depicts the six common part product life-cycle stages: introduction, growth, maturity, decline, phase-out, and obsolescence (also referred to as discontinuance) (Solomon, et al., 2000).[1] The Electronic Industries Alliance published a similar product life cycle model in EIA-724: "Product Life-Cycle Data Model" (EIA-724, 1997).

Product life cycle curve models are commonly used to describe the maturity of specific parts. For example, Freescale Semiconductor Inc. uses a more specific product life cycle stage model in which the stage "product introduction pending" has been added and the existing stage "growth" is divided into "product rapid growth" and "product stable growth." Moreover, the stage "product last shipments" is inserted between the stages "discontinuance" and "obsolescence" (Freescale, 2010a). In order to provide a common basis, the product life cycle model shown in Table 3-1 and Figure 3-2 will be used in this book.

Product life cycle stages have played an essential role since the first product portfolio analyses were used. For example, Boston Consulting Group described a useful and powerful model in 1970 called "The Product Portfolio" (Boston Consulting Group, 1970), which categorizes manufacturers' products into the four stages presented in the last row of Table 3-1.

[1]Several additional phases have been proposed, including introduction, pending (prior to introduction), and splitting the obsolescence stage into last shipment and discontinued or purged.

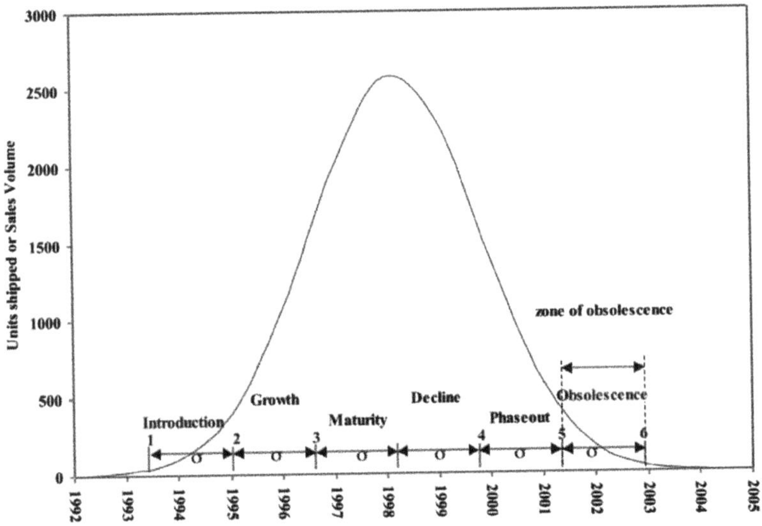

FIGURE 3-2 A standardized product life cycle curve.

3.1.1 Introduction Stage

New electronic parts introduced into the market may either be revolutionary or evolutionary. A part is considered "revolutionary if it carries out a new essential function or it uses a new semiconductor or packaging technology. Generally, a part is considered evolutionary if it is simply an improved version of some existing part, perhaps with added features or enhancements. The first 64K dynamic random access memory (DRAM) from IBM Microelectronics was an example of a revolutionary part. Current 64M DRAMs are examples of evolutionary parts. The latest revolutionary technology successfully introduced in the DRAM sector was the DDR3 SDRAM in 2006. Current evolutionary technology parts include low-power DRAMs to expand battery life in mobile devices, such as Mobile LPDDR2 (Micron, 2010b).

The introduction stage in a part's product life cycle is usually characterized by high production costs because of low yield, frequent modifications, low-production volumes, and lack of specialized production equipment. Marketing costs, at this stage, may also be high. Customers who buy a part in its introductory stage tend to value performance over price. Parts in the introduction stage can be associated with question mark status in the famous portfolio analysis by the Boston Consulting Group (Boston Consulting Group, 1970).

3.1.2 Growth Stage

The growth stage is characterized by the part's market acceptance. Increased sales during this stage may justify the use of specialized equipment for production, which in turn improves economies of scale of production.

TABLE 3-1 Typical Product Life Cycle Characteristics for the Six Generic Stages of a Part's Product Life Cycle

	Product Life-Cycle Stage					
Characteristic	Introduction	Growth	Maturity	Decline	Phase-out	Obsolescence
Sales	Slow but increasing	Increaszing rapidly	High	Decreasing	Lifetime buys may be offered	Sales only from aftermarket sources, if at all
Price	Highest	Declining	Low	Lowest	Low	Not applicable or very high if available from aftermarket sources
Usage	Low	Increasing	High	Decreasing	Decreasing	Low
Part modification	Periodic die shrinks and possible mask changes	Periodic die shrinks	Periodic die shrinks	Few or none	None	None
Manufacturing fab	High-tech—low volume	Standard—high volume	Standard—very high volume	Substandard—high volume	Outmoded—low volume	None or aftermarket fab
Competitors	Few	High	High	Declining	Declining	Few
Manufacturer profit	Low	Increasing	High	Decreasing	Decreasing	Decreasing
BCG–Matrix (Boston Consulting Group, 1970)	Question mark	Star	Cash cow	Poor dog	–	–

(Pecht and Das, 2000)

Mass production, mass distribution, and mass marketing often bring about price reductions. This stage often consists of the largest number of competitors, as opportunity-seeking firms are attracted by the part's profit potential. These competitors can be described as major companies with large investments in research and development. The equivalent in the Boston Consulting Group Matrix is the star, often also called the "rising star" (Boston Consulting Group, 1970), such as Mobile LPDDR memory (Micron, 2010a).

3.1.3 Maturity Stage

The maturity stage of the part's product life cycle is characterized by high-volume sales. Competitors with lower cost of production may enter the market, or domestic competitors may shift production facilities to less expensive locations to enable them to lower manufacturing costs. As of early 2010, Intel's notebook processors Intel Core™ 2 Duo are examples of mature parts. There are several other microprocessors with similar architecture and functionality available from other manufacturers that are generally sold at a lower price. Another example is DDR2 SDRAM, which is manufactured by numerous companies. Products at this stage are associated with the cash cow stage of the BCG matrix (Boston Consulting Group, 1970).

3.1.4 Decline Stage

Decreasing demand and generally decreasing profit margins characterize the decline stage. Toward the end of the decline stage, only a few specialized manufacturers generally remain in the market. Some manufacturers will sell their knowledge to other manufacturers at this stage. This "new" manufacturer will continue production while demand decreases or technology improves. Transistor-Transistor Logic (TTL) ICs are examples of parts that are available very late in this stage. Their continued but declining production is a result of their former widespread use in black and white televisions, and simple TTL logic ICs are still used in education settings. This product level is also called "poor dog" (Boston Consulting Group, 1970).

3.1.5 Phase-Out Stage

Phase-out occurs when the manufacturer decides on a date to stop production of the part. Generally, the manufacturer issues a discontinuance notice to customers, provides a last-time buy date, and suggests alternative parts or aftermarket manufacturers. As an example, on March 18, 2010, Intel Corporation announced the discontinuance of the Intel Xeon L3014, an embedded processor. Intel stated that the product would enter the discontinuance program on March 18, 2010 (Intel, 2010a) with the last (and non-cancelable) order date being March 18, 2011.

3.1.6 Discontinuance and Obsolescence

Discontinuance occurs when a manufacturer stops production of a part. In the beginning of 2010, the Intel Pentium M processor reached this stage. A part is obsolete when the technology that defines the part is no longer implemented. Thus, obsolescence occurs at the technology level, while discontinuance occurs at the part number or manufacturer-specific level. Diode transistor logic (DTL) and resistor transistor logic (RTL) parts are examples of obsolete part technologies. The National Semiconductor Military Quad SPST JFET analog switch in a ceramic DIP package is a discontinued part—the last-time buy date for this part was December 7, 1999 (National Semiconductor, 2002a). A non-military part of the same functionality and technology remains available from National Semiconductor. In this case, the military part is discontinued, but the technology is not obsolete.

3.2 SPECIAL CASES OF THE PRODUCT LIFE CYCLE CURVE

Not all parts conform to the six product life cycle stages presented here. Some parts undergo a false start and die out or may be associated with a niche market. Some parts may also be revitalized after the decline stage. Other possibilities can arise due to various economic, social, and environmental conditions.

A false start typically suggests that a part starts out with a period of strong growth, only to stall because of one or more of the following factors:

1. Introduction of a superior competing part.
2. Improvement of a competing part.
3. Identification of a problem associated with the part.
4. Failure to reach the critical mass that allows economies of scale to be realized.
5. Loss of a unique and compelling application for the part.

In 2000, Whirlpool and Cisco announced the first refrigerator with web access during the Consumer Electronics Show (Cisco, 2000). This refrigerator is an example of a false start, and as of 2010 web-accessible refrigerators are still not widespread.

Niche parts generally have some unique applications and thus have a relatively low but steady sales level. An example is GaAs ICs, which have found a niche market in telecommunications, military, and space applications.

Decline can often be delayed or reversed by revitalizing the original part. Defining new market segments, new applications, or creating a new image for the part and thereby increasing the demand can cause revitalization.

3.3 PRODUCT LIFE CYCLE STAGES AS A BASIS FOR FORECASTING

The product life cycle stage that a part is in can be used as a basis for forecasting the obsolescence date for a part. A product's current procurement status and its future development can be described by the product life cycle stage that it is in and through the use of changes in its product life cycle status. The product life cycle stage concept can also be applied to product families, companies, composite products, and technologies.

Chapter 4

Obsolescence Forecasting Methodologies

There are various technological forecasting methods such as Delphi, forecasting by analogy, growth curves, trend extrapolation, measures of technology, correlation methods, casual models, and combining forecasts (Martino, 1983). These methods are dominated by the "exploratory" attempt to estimate the future based on facts about the present and past. Alternative methods, known as normative methods (Martino, 1983), focus on future needs and track them backwards to the present.

All exploratory methods offer individual advantages and disadvantages—a discussion of which is out of the scope of this chapter—but the best forecasting results can only be achieved by a specific method adapted to an individual problem. To satisfy the needs for electronic part obsolescence forecasting, a combination of exploratory methods is generally used. The obsolescence forecasting methods discussed in this chapter are designed to be objective and reproducible so that they can be implemented in software in order to support forecasting for the large volume of components in electronic systems.

Taking a deeper look into obsolescence forecasting methodology, there are several traditional methods, such as the "scorecard" or the "availability factor" techniques. In the scorecard approach, the life cycle stage of the parts is determined from an array of technological attributes, and each attribute is given a life cycle code ranging from 1 through 5 (or 6 in some cases) and a corresponding weight. The overall life cycle stage for the part is determined by computing a simple weighted average of the life cycle codes for the attributes.

Strategies to the Prediction, Mitigation and Management of Product Obsolescence, First Edition. Bjoern Bartels, Ulrich Ermel, Michael Pecht and Peter Sandborn.

The availability factor method projects a "safe" usage window for a part. This approach uses market and technology factors to predict the obsolescence of parts with similar technology and market characteristics.

The availability method is mainly driven by market behavior without the life cycle model, whereas the scorecard method derives from evaluating the technical parameters (such as packaging style or process densities) without capturing market trends (Pecht et al., 2002).

To mitigate the disadvantages of each traditional approach, an alternative approach based on life cycle curve forecasting that combines the advantages of both traditional practices is described in Section 4.1. For parts that have no clear evolutionary driver, a procurement life forecasting approach is introduced in Section 4.2. Finally, a non-database method is presented in Section 4.3.

4.1 OBSOLESCENCE FORECASTING—PARTS WITH EVOLUTIONARY PARAMETRIC DRIVERS

This section presents forecasting models that are applicable to electronic parts that have clearly identifiable evolutionary parametric drivers. An evolutionary parametric driver is a parameter (or a combination of parameters) describing a part that evolves over time. For example, for flash memory chips, an evolutionary parametric driver is memory size, whereas for microprocessors it has traditionally been clock frequency (although recently this has begun to give way to power consumption). Section 4.1.1 describes the basic life cycle curve forecasting method, and Section 4.1.2 describes the extension of the life cycle curve forecasting method that uses data mining to determine the window of obsolescence of the parts.

4.1.1 Basic Life Cycle Curve Forecasting Method (Solomon et al., 2000)

This method is based upon the life cycle curve description of an electronic part discussed in Chapter 3. In Figure 4-1, the light efficiency of various types of lamps is plotted versus time. The light efficiency is a clear evolutionary parametric driver for lamps. The maturity of each lamp technology is clearly correlated to the appearance of a more efficient lamp technology. Notice that technologies "overlap,": the introduction of new technology does not generally lead to the predecessor's immediate obsolescence. The obsolescence forecasting methodology described in this section is based on predicting the life cycle curves of technologies that are driven by parameters that clearly evolve in time, similar to the lamps shown in Figure 4-1. This method is composed of the six steps shown in Figure 4-2.

4.1.1.1 Step 1: Identify Part/Technology Group First, the device's material group has to be identified. The material group describes the general class of functionality. For example, a material group "memories" would refer to the

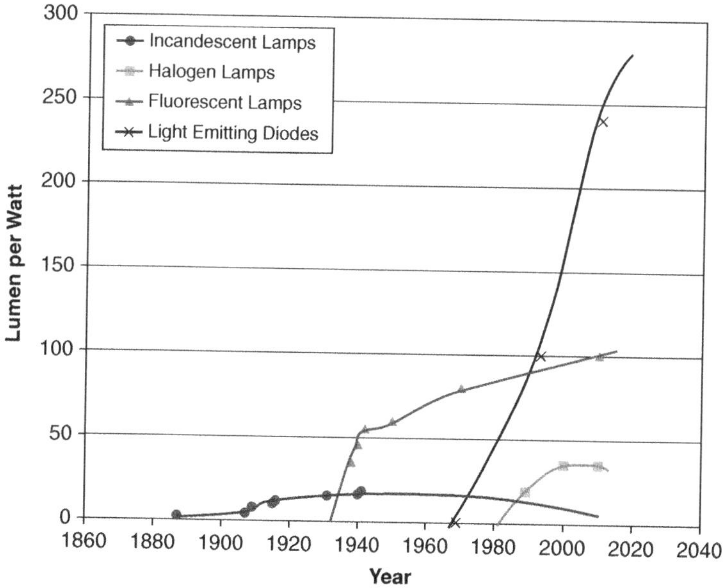

FIGURE 4-1 Lumen per watt for incandescent lamps, halogen lamps, fluorescent lamps, and light emitting diodes. Adapted and updated in 2010 from (Martino, 1983).

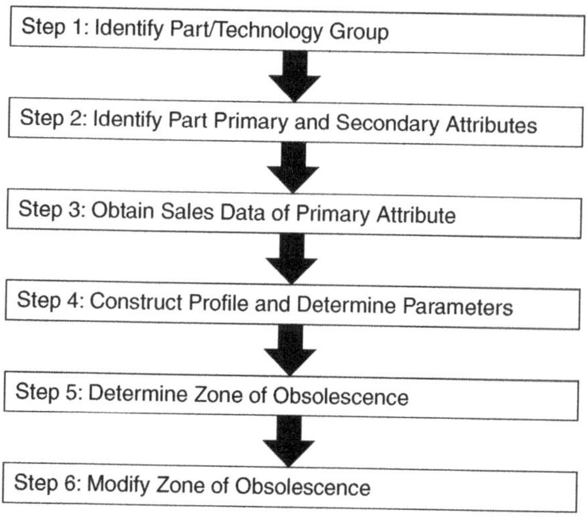

FIGURE 4-2 Basic life cycle curve forecasting method.

entirety of all memory components, including DRAM, Flash, EEPROM, or any other memory technology.

Once the material group is identified, the technology group is identified. The technology group summarizes all parts based on the same technology characteristics. For instance, a group could consist of all 256M NOR-Flash memories with 64-pin FBGA packages, 3.0V supply voltage, and page mode, regardless of their manufacturers.

Consider a single-level BOM that contains a capacitor, for example, the VJ0805A1R8KXXAT68 from Vishay (Vishay, 2010). The material group would be defined as capacitor. The technology group would characterize the set union of capacitors with 1.8pF high reliability, 0805 case, 10 percent tolerance, 100 percent tin-plated, and 25V DC rating. The Kemet capacitor C0805Q189K3XAC would be another member of this technological group (Kemet, 2010). Obtaining the technology group is important for the next step in the methodology.

4.1.1.2 *Step 2: Identify the Part's Primary and Secondary Attributes* This step involves the identification of a part's primary attributes. Primary attributes are properties that describe the part's technology group. According to Pecht (2002), a primary attribute is a characteristic that defines a part/technology group. For example, the primary attribute for a memory part is memory density. The memory density is an evolutionary parametric driver (a parametric value whose evolution over time is clearly correlated to the part's introduction and obsolescence).

Identifying the primary attribute is not always simple. For instance, Micron Technologies offers the same NOR-Flash P30/P33 on 65nm technology with different supply voltages and densities in order to meet the requirements for mobile devices as well as traditional requirements (Micron, 2010c). In order to constantly reduce supply voltage and current, the evolutional driver might develop toward power consumption instead of memory density. Because most parts cannot generally be reduced to one single attribute, other attributes will be included in the forecasting methodology as well. Finally, it is reasonable to decide on the factor with the strongest evolutionary influence on the part as being the part's primary attribute, as shown in Table 4-1.

A secondary attribute is a characteristic of the part/technology group that can modify the range of years to obsolescence and the life cycle stage of a part/technology group (Pecht et al., 2002). For instance, chip packaging, supply voltages, and access methods are typical secondary attributes.

These attributes have an immense impact on the life cycle curve forecasting model, because even if the density of a memory chip is not obsolete, a design may require the chip in a package that is obsolete. This means that in certain cases, secondary attributes are valued over primary attributes.

Table 4-1 refers to common primary and secondary attributes for various IC part classes. It is recommended not to exceed the number of three secondary attributes.

If a primary attribute for the part type cannot be determined, or if the primary attribute is not an evolutionary parametric driver, then the procurement life forecasting approach described in Section 4.2 should be used.

TABLE 4-1 Primary and Secondary Attributes for IC Part Classes

Part Class	Primary Attribute	Secondary Attribute(s)
DRAM	Memory size	DRAM type (EDO, SDRAM, DDR, DDR2, DDR3, GDDR5, other types[1]) Package style (TSOP, BGA, FBGA, VFBGA, other package styles[2]) Voltage (3.3V, 1.8V, 1.5V, <1.5V, and so on)
SRAM	Memory size	SRAM type (such as synchronous, asynchronous, QDRI/II/III, DDR II, SB & SPB, NtRAM (FT & PP, DLW) Package style (such as TQFP, FBA, BGA, SOJ, TSOP, other package styles[2]) Voltage (5V, 3.3V, 2.5V, 1.8V, 1.65V–3.6V, and so on)
Flash	Memory size	Flash type (NOR, NAND, Strata Flash, NROM) Package style (TSOP, FBGA, PLCC, VFBGA, other package styles-[2]) Voltage (5V, 3.0V, 2.5V, 1.8V, and so on)
EPROM	Memory size	EPROM type (such as OTP) Package style (PLCC, TSOP, FDIP, PDIP, other package styles[2]) Voltage (5V, 3V, 2.7V, and so on)
EEPROM	Memory size	EEPROM types (Parallel, Serial) Package style (PLCC, TSOP, PDIP, other package styles[2]) Voltage (4.5V–5V, 2.5–5.5V, 1.8–5.5V)
Micro-controllers and DSP	Data bus width	Package style (FCBGA, NFBGA, LQFP, SOIC, MAPGA, other package styles[2]) Voltage (5V, 3.3V, 3V, 2.7V, 2.5V, 1.8V, and so on)
Micro-processor	Data bus width	Package style (FCBGA, NFBGA, LQFP, SOIC, MAPGA, other package styles[2]) Voltage (2.5V, 1.8V, 1.3V, 1.2V, 1.0V, and so on) Frequency

(Continued)

[1] Including manufacturer-specific low-power types for mobile applications (such as ISSI's Mobile DDR, Micron's Mobile LPDRAM, or Elpida's XDR DRAM).
[2] Including older package styles such as SOP, DIP, and other manufacturer-specific packages.

TABLE 4-1 Continued

Part Class	Primary Attribute	Secondary Attribute(s)
Logic parts	Logic family (Advanced-Low-Power-Shottky-TTL, Advanced-Low-Voltage-CMOS)	Package style (DIP, SOIC, TSSOP, other package styles[2]) Voltage (5V, 3–5V, 3–3.5V, 1.2–3.6V, and so on)
Analog parts	Comparator	Package style (LFCDP, SOIC, TSSOP, DIP, PLCC, other package styles[2])
Analog parts	Voltage regulator	Package style (WDFN, SOT, SOIC, TSOT, DFN, other package styles[2]) Output current
Analog to Digital	Resolution	Package style (TSSOP, SOIC, PDIP, other package styles[2]) Sample rate
Interfaces	Interface	Package style (BGA, LQFP, PDIP, SOIC, TSSOP other package styles[2]) Voltage (5V, 3.3V, 1.95V, 1.5V, and so on)
	Other analog	Package style (LFCDP, SOIC, TSSOP, DIP, other package styles[2]) Voltage (5V, 11V, 12V, 15V, 30V, and so on)
ASICs	CMOS (HiPerMOS, RDCMOS) SOI	Package style (BGA, SOIC, TSSOP, DIP, other package styles[2]) Voltage (3–5V, 3–3.5V, 1–2V, and so on)

(Pecht, et al., 2002).

4.1.1.3 *Step 3: Obtain Sales Data Associated with the Primary Attribute*

In this step, sales data are collected in terms of the primary attribute. In theory, sales data are a direct indicator of the life cycle stage of the primary attribute. The sales data may be in the form of the number of units shipped. If unit sales data are not available, sales in market dollars or percentage market share may be used, if the total market does not increase appreciably over time. For instance, these sales data are available from market research organizations and may be available from manufacturers and distributors directly. Sales data can be influenced by general economic factors. If no sales data are available, the procurement life model described in Section 4.2 is recommended.

4.1.1.4 *Step 4: Construct the Life Cycle Curve and Determine Parameters*

The life cycle curve is constructed by fitting the sales data of the primary attribute to a life cycle forecasting distribution. Each life cycle stage is defined in terms of its distance from the mean (μ) measured in standard deviations (σ) when the life cycle curve is fit with a Gaussian form. The zone of obsolescence is defined as the ordered pair, ($\mu + 2.5\,\sigma - p$, $\mu + 3.5\,\sigma - p$), where p is the date of assessment. This ordered pair gives the number of years from the present to the beginning and end of the window of obsolescence. Figure 4-3 shows the life cycle curve of the 16Mb DRAM (Pecht et al., 2002).

Gaussian distributions have been used by the Electronic Industries Association (EIA) in their EIA724 standardized product life cycle curve and hence are well known and familiar to equipment manufacturers (EIA-724, 1997). The equation of the life cycle curve (fit with a Gaussian function) is:

$$f(x) = ke^{-\frac{(x-\mu)^2}{2\sigma^2}} \tag{4-1}$$

The function $f(x)$ is a trend equation that gives the relation of the life cycle curve parameters (μ and σ for Gaussian functions) with the primary attribute. Function $f(x)$ is defined by the mean μ, the point in time of the sales peak of the curve, the standard deviation σ, and the year x. The factor k is the sales peak, the number of units shipped, or the market share.

Figure 4-4 shows a sample trend curve for μ versus DRAM memory size. The standard deviation, σ, can also be computed from the curve fits. Figure 4-5 shows a sample trend curve for σ versus DRAM memory size trend.

From the curves shown in Figures 4-4 and 4-5, the year in which the peak sales will occur and the width of the distribution for new generations of parts sharing the same primary attribute can be forecasted.

4.1.1.5 Step 5: Determine the Zone of Obsolescence The zone of obsolescence refers to a period of time in which the part/technology group has a high probability of becoming obsolete. The ordered pair $(\mu + 2.5\sigma - p,$

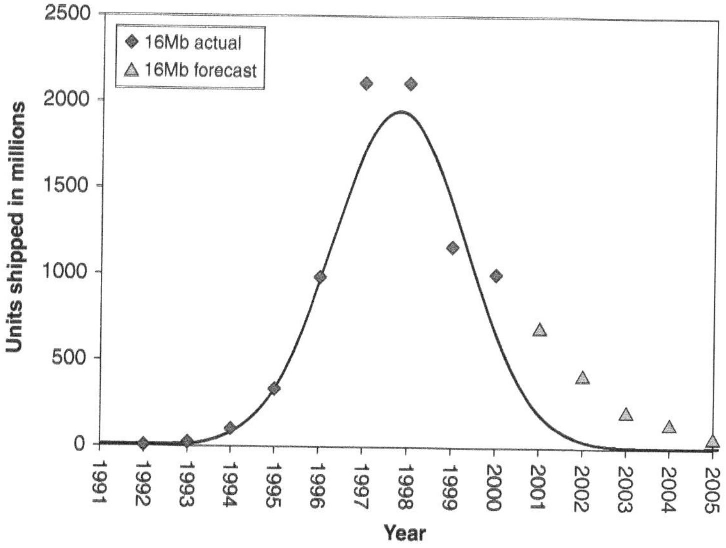

FIGURE 4-3 Life cycle curve of a 16Mb DRAM.

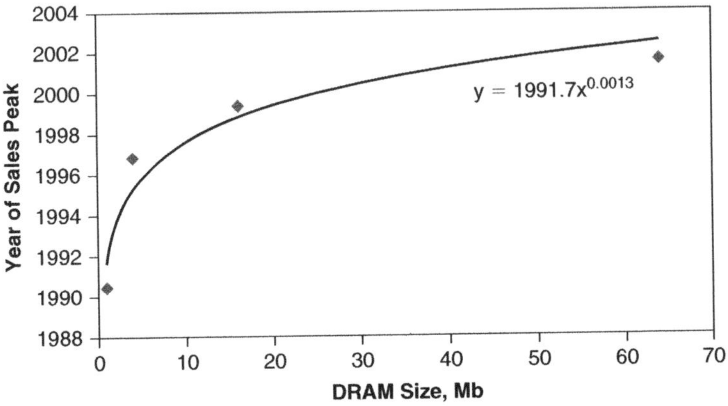

FIGURE 4-4 Mean versus DRAM size.

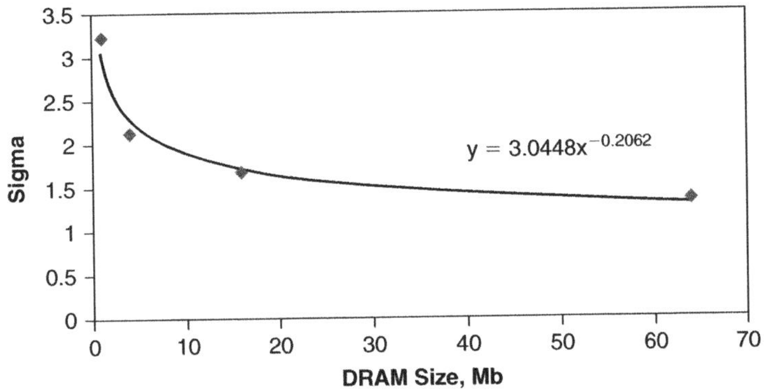

FIGURE 4-5 Standard deviation versus DRAM size.

$\mu + 3.5\sigma - p$) gives the zone of obsolescence, where p is the date of assessment. For the 16Mb DRAM shown in Figure 4-6, the zone of obsolescence is 2.7 to 4.3 years. Life cycle stages are determined by dividing the life cycle curve for the primary attribute into the basic life cycle stages (see Chapter 3 and Table 4-2).

4.1.1.6 Step 6: Modify the Zone of Obsolescence The life cycle curve of the part/technology group may require modification by secondary attributes. If the years to obsolescence for any of the secondary attributes fall within the life span ($\pm 3\sigma$ years) of the main attribute, the years to obsolescence for the generic part must be modified. The algorithm is summarized in Table 4-3, where the secondary attributes are denoted by j, with μ_j and σ_j referring to the mean and standard deviation of the secondary attribute.

Table 4-4 depicts the application of the algorithm to evaluate the zone of obsolescence (the span of time during which the part is most likely to be

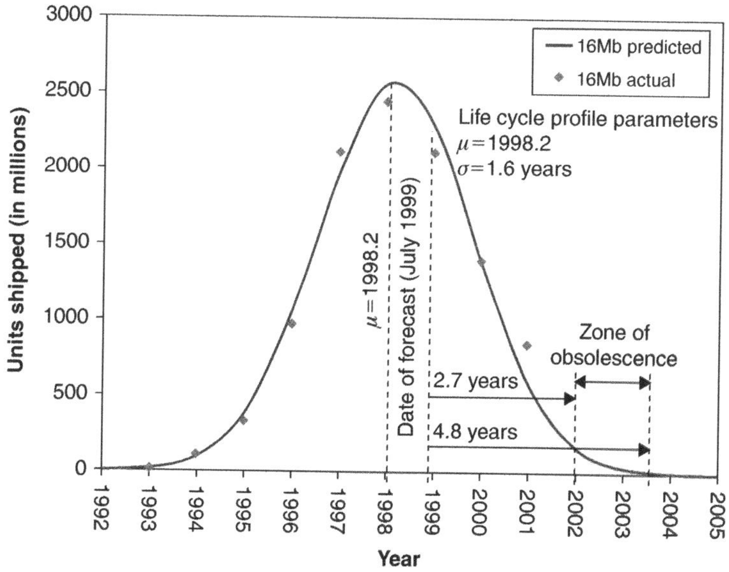

FIGURE 4-6 Zone of obsolescence and life cycle curve for a 16Mb DRAM.

TABLE 4-2 Life Cycle Stages and Definition of Life Cycle Zones

Life Cycle Stage Name	Life Cycle Zones $(x_{i,j}(\mu, \sigma),\ x_{i,j+1}(\mu, \sigma))$
Introduction	$(\mu - 3\sigma,\ \mu - 2\sigma)$
Growth	$(\mu - 2\ \sigma,\ \mu - \sigma)$
Maturity	$(\mu - \sigma,\ \mu + \sigma)$
Decline	$(\mu + \sigma,\ \mu + 2\sigma)$
Phase out	$(\mu + 2\sigma,\ \mu + 3\sigma)$
Obsolescence	$(\mu + 2.5\sigma,\ \mu + 3.5\sigma)$

TABLE 4-3 Algorithm for Choosing a Zone for the Years to Obsolescence from Secondary Attributes (p = Date of Assessment)

No.	Condition	Modified zone of obsolescence (a, b)
1	$\mu_j + 3.5\sigma_j - p < \mu + 2.5\sigma - p$	$\mu_j + 2.5\ \sigma_j - p,\ \mu_j + 3.5\sigma_j - p$
2	$\mu + 2.5\sigma - p \leq \mu_j + 3.5\sigma_j - p \leq \mu + 3.5\sigma - p$	$min\ [\mu + 2.5\sigma - p,\ \mu_j + 2.5\sigma_j - p]$, $min\ [\mu + 3.5\sigma - p,\ \mu_j + 3.5\sigma_j - p]$
3	$\mu_j + 3.5\sigma_j - p > \mu + 3.5\sigma - p$	$min\ [\mu_j + 2.5\sigma_j - p,\ \mu + 2.5\sigma - p]$, $\mu + 3.5\sigma - p$

TABLE 4-4 Case Study: Modified Zone of Obsolescence of a 16Mb, EDO, 5V, SOP DRAM

Attribute	Zone of obsolescence (years)	Modified zone of obsolescence (years)
Memory density (16Mb)	(2.7, 4.3)	–
DRAM type (EDO)	(2.2, 3.8)	–
Voltage (5V)	(1.3, 5.3)	(1.3, 3.8)
Package style (SOP)	(13.7, 18.9)	–

discontinued) of a 16Mb, EDO, 5V, SOP DRAM. Table 4-4 shows the shift in the life cycle profile of the 16Mb DRAM resulting from the two secondary attributes. The curve fit is based on the trend equation $\sigma = 3.1M^{-0.23}$ where σ is the standard deviation and M is the corresponding memory size. Thus, the years to obsolescence of a 16Mb DRAM decreases from 2.7 to 1.3 years due to effect of secondary attributes.

4.1.1.7 Summary An electronic part usually advances through six stages: introduction, growth, maturity, decline, phase-out, and obsolescence. The part life cycle curve provides a basis for part analysis and forecasts.

The life cycle forecasting model presented achieves the following:

1. Captures market trends by identifying a set of quantifiable market and technological attributes (such as memory density, part supply voltage, memory part type, package style) that govern the growth and demise of part/technology groups.
2. Computes both years to obsolescence and life cycle stage based on statistical analysis of sales data for the market and technology attributes for the part/technology group.
3. Computes an overall risk factor associated with a specific part number by implementing market factors, such as component risk, manufacturer market share, and part life cycle information.

The impact of the aftermarket is not accounted for in the life cycle curve approach. Aftermarket sources may continue to manufacture the part after the original manufacturers have made obsolete the actual part/technology groups. Equipment suppliers may build "special" relationships with aftermarket sources, sunset distributors, and emulation sources to ensure continued availability.

Some market factors are manufacturer- and application-specific, which should be considered an additional risk associated with using a part. Any high-volume end-application that experiences growth encourages demand for the parts that it is composed of. This demand reduces the risk of suppliers leaving the market, but may accelerate the evolution of parts and thereby decrease part procurement lives (that is, accelerate obsolescence). For example, a strong growth in the demand for cellular phones has led to strong growth in the flash memory and EEPROM markets. Manufacturers having significant market share and profitability have a reduced probability of leaving the market. For example, it is highly improbable that Intel will quit the microprocessor market, as it controlled an 80.7 percent market share as of the second quarter of 2010 (Heise, 2010a), and microprocessors represent Intel's core competency.

The number of sources for a part does not necessarily indicate health. For example, the 5V 256K asynchronous SRAM in a DIP package was manufactured by nine manufacturers in 2001. However, the part/technology group was made obsolete a few years later by manufacturers who were replacing the part with SRAMs of the same functionality but in newer package styles and lower voltages.

The fact that there are only a few sources of parts may suggest that the manufacturers still in business command most of the market share. For example, some aftermarket manufacturers continue to manufacture families of TTL logic even after the original manufacturers have discontinued their product lines. A big market player quitting the business does not necessarily mean the death of the part/technology group. Manufacturers may decide to discontinue a product line for a host of business reasons that may not have much to do with part/technology obsolescence. This occurrence is especially true in the volatile memory market. For example, in September 1998, Texas Instruments decided to sell off its entire memory line to Micron and focus on digital signal processors (DSPs). This does not mean that the memory market is dying out; in fact, this exemplifies the market's vitality.

4.1.2 Advanced Life Cycle Curve Method (Sandborn et al., 2007a)

To improve the basic life cycle curve forecasting approach described in Section 4.1.1, an advanced life cycle curve approach was developed. This method is a combination of life cycle curve forecasting and the determination of electronic part vendor-specific windows of obsolescence using data mining of historical last-order or last-ship dates. The methodology not only enables more accurate obsolescence forecasts but can also generate forecasts for user-specified confidence levels. The methodology described in this section is demonstrated on both individual parts and modules.

The obsolescence forecasting approach discussed in this section is an extension of the life cycle curve forecasting methodology based on curve-fitting sales data for an electronic part described in Section 4.1.1. In the previously discussed methodology, attributes of the curve fits (such as mean and standard deviation for sales data fitted with a Gaussian distribution) are plotted and trend equations are created that can be used for forecasting the life cycle curve of future versions of the part type. This obsolescence forecasting approach uses a fixed "window of obsolescence" determined by a fixed number of standard deviations from the peak sales year of the parts categorized by primary and secondary attributes. In the original baseline methodology, the window of obsolescence specification was defined to be at 2.5σ to 3.5σ after the peak sales date (μ) and possibly updated using secondary attributes. In reality, the window of obsolescence specification is not a constant, but depends on numerous factors.

4.1.2.1 *Determining the Window of Obsolescence via Data Mining* In actuality, the window of obsolescence specification is dependent on manufacturer-specific and part-specific business practices (Sandborn et al., 2007a). For a particular part type (such as flash memory), historical last-order date data are collected and sorted by manufacturer.[3] Each part instance (data entry) in the resulting sorted data has a specific value of primary attribute (such as 32Mb) for which the peak sales date μ_p and standard deviation σ_p can be computed using the previously created trend equations.[4]

The last order date for the part instance is then normalized relative to the peak sales year (that is, the last order date is expressed as a number of standard deviations after the peak sales year). The normalization is performed for every part instance for the selected part type and manufacturer.

Next, a histogram of the normalized vendor-specific last order dates is plotted. The histogram represents a probability distribution of when (relative to the peak sales year) the specific manufacturer discontinues the part type. As an example, Figure 4-7 shows the histogram for Atmel (ATM) flash memory (based on 57 last order dates mined from PartMiner CAPS Expert). To quantify the manufacturer-specific obsolescence probability, the histogram is fitted with a Gaussian function and the parameters of the fit are extracted: μ_{lo} and σ_{lo}. The window of obsolescence specification is then given by

$$\text{Obsolescence window} = \mu_p + (\mu_{lo} \pm x\sigma_{lo})\sigma_p \tag{4-2}$$

where x depends on the confidence level desired, according, to Table 4-5.

The parameters μ_{lo} and σ_{lo} were generated from the ATM flash memory by finding the mean and standard deviation of the histogram shown in Figure 4-7. By combining the life cycle curve trends and the ATM-specific obsolescence window, the resulting obsolescence dates for ATM flash memory are given by

[3] The last order date is the last date that a manufacturer will accept an order for a part. After the last order date has passed, the part is considered to be no longer purchasable. Once the last time shipment has passed, the part will no longer be manufactured by the original source and is considered obsolete. The obsolescence of a part does not necessarily correlate to the part's availability; some parts remain available through aftermarket sources and brokers for considerable periods of time after the original manufacturer has discontinued them.

[4] For flash memory these are given by

$$\mu_p = 1.5663\ln(M) + 1997.2$$

$$\sigma_p = -0.0281\ln(M) + 2.2479$$

where M is the size of the flash memory chip in Mb.

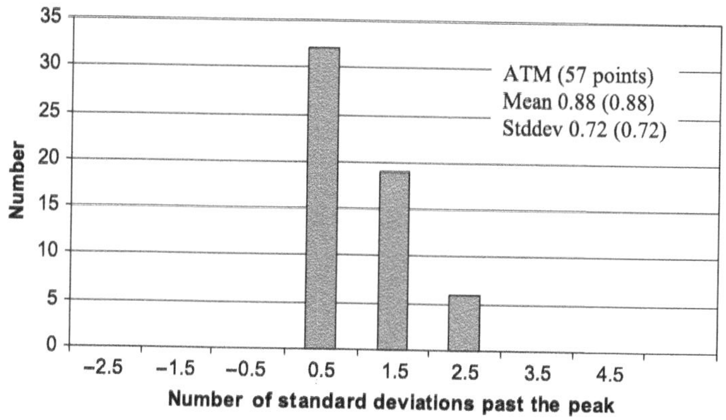

FIGURE 4-7 Atmel (ATM) flash memory last order dates projected in intervals.

TABLE 4-5 Confidence Level in Relation to Factor *x* (For instance to get a confidence of 99.7%, a value of 3 has to be chosen for *x* in Eq. 4-2)

Confidence Level	Factor *x*
68.3%	1
86.6%	1.5
91.1%	1.7
95.5%	2
97.9%	2.3
99.7%	3

(Bartsch, 1994).

Eq. 4-3 is a function of the size in megabits (M) and confidence level desired:

$$\text{Obsolescence date} = \underbrace{1.5663\ln(M) + 1997.2}_{\text{peak sales date}} + \overbrace{[0.88 \pm 0.72x]}^{\substack{\text{Number of Standard deviations}\\ \text{past the peak for ATM Flash}}} \underbrace{(-0.0281\ln(M) + 2.2479)}_{\substack{\text{Standard deviation}\\ \text{in sales data}}} \quad (4\text{-}3)$$

Eq. 4-3 assumes that the uncertainty in the window of obsolescence dominates the model uncertainty associated with the trend equations. Using the methodology for the entire set of flash memory provided by PartMiner (262 data points) yields the results shown in Figure 4-8. The diagonal line in the plot represents exact agreement between the prediction and the actual results. The

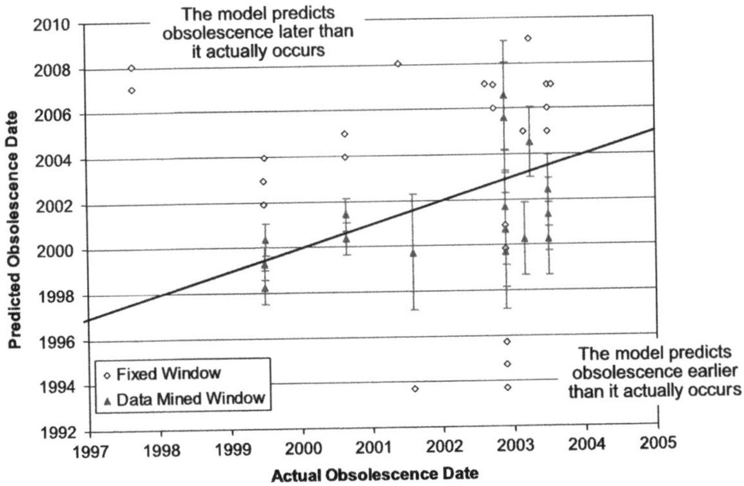

FIGURE 4-8 Plot of forecasting results for 262 flash memory chips. Fixed window model assumes a fixed window of obsolescence specification of 2.5 σ_p to 3.5 σ_p; data-mined window model assumes the manufacturer-specified window of obsolescence.

error bars represent a 68.3 percent confidence level. The accuracy with which the improved algorithm forecasts the obsolescence of parts is a substantial improvement over the original algorithm (basic life cycle method with a fixed window).

4.1.2.2 Application of Data Mining Determined Windows of Obsolescence to Memory Modules As a further demonstration of the methodology described in this section, consider its application to memory modules that are made up of multiple chips. The obsolescence of memory modules is not generally dictated by the obsolescence of the memory chips that are embedded within them. Rather, the obsolescence of memory modules is related to the beginning of the availability of monolithic replacements for identical amounts of memory. As an example, in Figure 4-9, the 16Mb DRAM module became obsolete when monolithic 16Mb DRAM chips became available.

In the case of DRAM memory modules, the last-order date data is collected. Each module instance (data entry) has a specific value for its primary attribute (such as 16Mb). For each module instance, the peak sales date μ_p and standard deviation σ_p are computed for the monolithic equivalent. The last order date for the module instance is then mapped (normalized) to the standard deviations before the peak sales date for the monolithic equivalent. In the case of memory modules, there was no need to sort the data by vendor, since all the vendors considered appear to be discontinuing memory modules based on the same driver. Figure 4-10 shows a curve fit of the resulting data-mined last-order dates mapped to the life cycle curve of the monolithic equivalents.

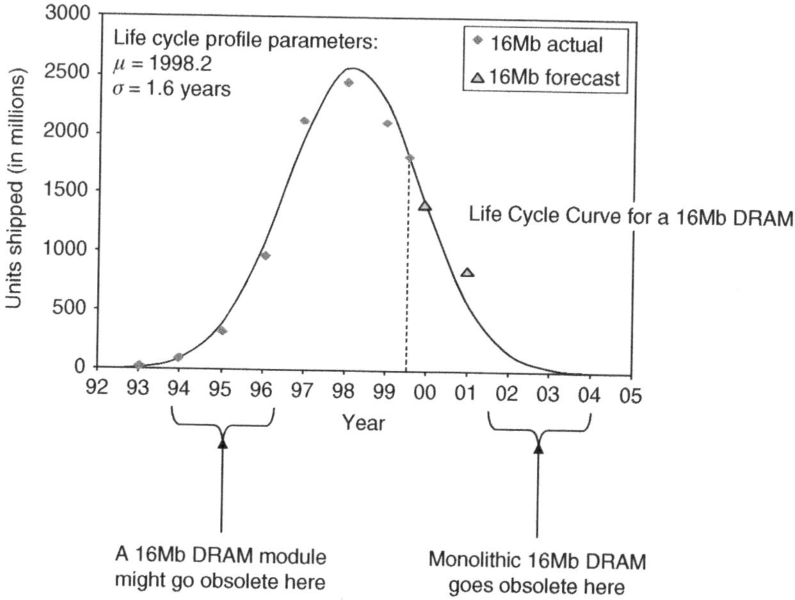

FIGURE 4-9 Obsolescence characteristics of DRAM memory modules versus monolithic DRAM.

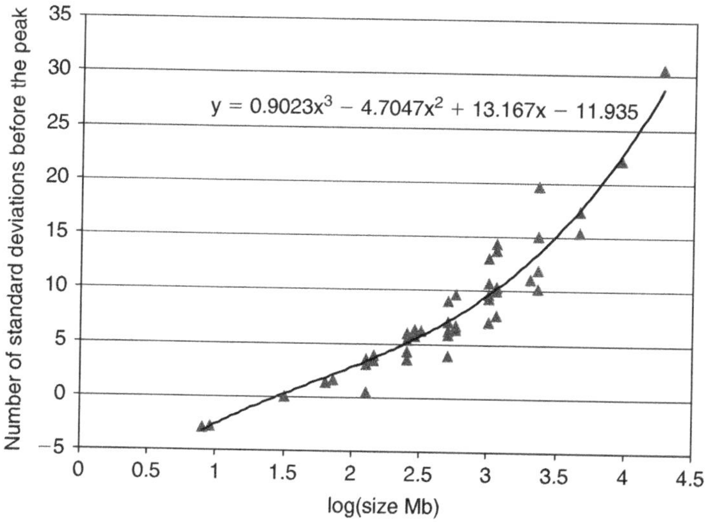

FIGURE 4-10 Data-mined data mapping for DRAM memory modules. Number of standard deviations before the peak versus the logarithm of the memory module size.

Armed with the relation shown in Figure 4-10 and the life cycle curve trends for DRAMs (see Figure 4-6), obsolescence dates for DRAM memory modules are given by Eq. 4-4:

$$= \underbrace{1991.8M^{-0.0011}}_{\substack{\text{Peak sales date of} \\ \textit{monolithic}\text{ DRAM}}} - \overbrace{[0.9023x^3 - 4.7047x^2 + 13.167x - 11.935]}^{\substack{\text{Number of Standard deviations before the} \\ \text{peak of an equivalent } \textit{monolithic}\text{ DRAM}}} \underbrace{(3.1M^{-0.23})}_{\substack{\text{Standard deviation} \\ \text{in sales data for the} \\ \textit{monolithic}\text{ DRAM}}} \tag{4-4}$$

In Eq. 4-4, $x = log(M) \pm \alpha$, where the value of α depends on what confidence level is desired (for example, $\alpha = 0$ gives you the curve fit in Figure 4-10; $\alpha = 0.3$ gives ~90% confidence level).

To summarize, the advanced life cycle method described in this section extends the basic life cycle method using manufacturer-specific product marketing behavior. The advanced method allows analysis of obsolescence for piece parts and modules. Moreover, it is possible to specify the required level of confidence in the forecast. However, this approach requires a large amount of data in order to reduce statistical uncertainties. Last, to use either the basic or advanced methods described so far in this chapter, it is necessary to identify the part's evolutional parametric drivers (memory, density, voltage, packaging, and so on). In order to treat parts with no clear evolutionary parametric driver, a procurement life forecasting method has been developed and will be discussed in the next section.

4.2 OBSOLESCENCE FORECASTING—PARTS WITHOUT EVOLUTIONARY PARAMETRIC DRIVERS (SANDBORN ET AL., 2011)

This section introduces a methodology for generating algorithms that can be used to predict the obsolescence dates for electronic parts that do not have clear evolutionary parametric drivers. The method is based on the calculation of procurement lifetime using databases of previous obsolescence events and introduced parts that have not gone obsolete. The methodology is demonstrated on a range of different discrete semiconductor electronic parts for the trending of specific part attributes.

4.2.1 Procurement Lifetime

The previously discussed data mining method for forecasting obsolescence has been shown to work well when there are identifiable evolutionary parametric drivers (Section 4.1). An evolutionary parametric driver is a parameter (or a combination of parameters) describing how the part evolves over time. For example, for flash memory chips an evolutionary parametric driver is memory size; traditionally for microprocessors it has been clock frequency (although recently this has begun to give way to power consumption). Unfortunately, for the majority of electronic parts, there is no simple evolutionary parametric driver that can be identified, and previously proposed data mining approaches cannot be used (diodes, transistors, operational amplifiers, multiplexer, buffers, and so on).

In this section, a methodology for formulating obsolescence forecasting algorithms based on predicting the part's procurement life is presented that does not depend on the identification of an evolutionary parametric driver for the part (Sandborn et al., 2011). The procurement life for a part is defined as,

$$L_P = D_O - D_I \tag{4-5}$$

where

L_P = Procurement life, the amount of time the part was (or will be) available for procurement from its original manufacturer.

D_O = Obsolescence date, the date that the original manufacturer discontinued or will discontinue the part.

D_I = Introduction date, the date that the original manufacturer introduced the part.

The concept of procurement life has also been referred to as "product lifetime" (Bayus, 1998) and "duration time" in the marketing literature (Helsen and Schmittlein, 1993). In this section, we are interested in exploring the correlation between procurement lifetime and the introduction date for electronic parts. Two specific results are of interest for forecasting uses: first, the mean procurement lifetime as a function of introduction date and, second, the effective worst-case procurement lifetime as a function of introduction date.

4.2.2 Electronic Part Introduction Date and Obsolescence Date Data

A part database from SiliconExpert was used for the analysis in this section. The part database was created and is maintained using parts data that is sourced via web crawling of manufacturers' websites for revision control and through direct relationships with manufacturers in which feeds of component data are supplied on a weekly, monthly, or quarterly basis.

The frequency of updates is dependent upon how often datasheets and product information are revised by document controllers internally at each manufacturer. In addition, data feeds provided via direct relationships enlist datasheets and parametric information for new products introduced to the market. Organization of the database is done through a strict taxonomy. Revisions to the taxonomy occur only for expansion and contraction of existing product lines to ensure valid differentiation between various subcategories. As

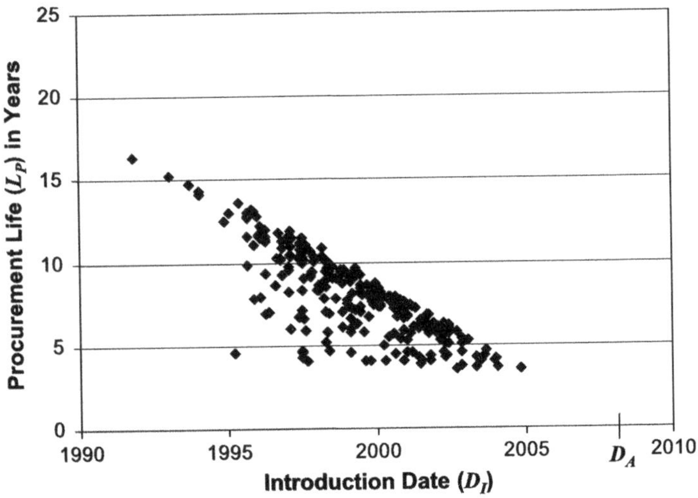

FIGURE 4-11 347 obsolete linear regulators from 33 manufacturers. $D_A = 2008$, the analysis date (the date on which the analysis was performed).

of June 14, 2010, the database contained more than 157 million electronic parts spanning 337 product lines from 11,054 manufacturers.

Figure 4-11 shows an example plot of procurement life versus introduction date for obsolete linear regulators (a common electronic part that regulates voltage between the supply and the load and provides a constant voltage by varying its effective resistance) mined from the SiliconExpert database.

4.2.3 Determining Mean Procurement Lifetimes

The mean procurement lifetimes for parts can be analyzed using the statistical framework for failure time analysis (Bayus, 1998). This approach has been previously used to determine the mean product life cycle lengths for personal computers (Bayus, 1998). The approach in Bayus (1998) and Helsen and Schmittlein (1993), however, has never been applied to the forecasting of obsolescence or to the procurement of electronic parts.

The event of interest in this section is the discontinuance (obsolescence) of a part. The data used includes the introduction dates of all the parts of a particular type and the obsolescence dates for the parts that have occurred up to 2008. An obsolescence event is not observed for every part in the data set, since some of the introduced parts had not gone obsolete as of the analysis date (the observations are right censored).

Following the analysis method in Helsen and Schmittlein (1993) and representing the data for the linear regulator example as a distribution of procurement lifetimes (event density), $f(t)$, with a corresponding cumulative distribution function, $F(t)$, the hazard rate, $h(t)$ is given by,

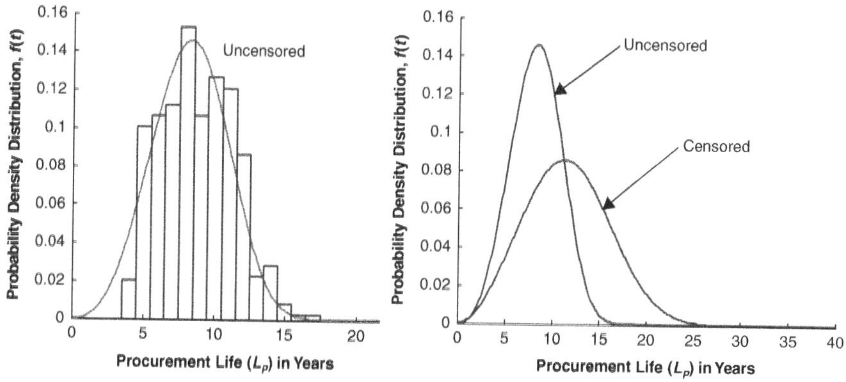

FIGURE 4-12 The distribution of procurement lifetimes for linear regulators. The histogram on the left side corresponds to the data in Figure 4-11. The mean procurement lifetime (censored) = 11.63 years, $\beta = 2.84$, $\eta = 13.06$. The parameters are based on a maximum likelihood estimate (MLE) using a two-parameter Weibull fit.

$$h(t) = \frac{f(t)}{1 - F(t)} \tag{4-6}$$

The hazard rate is the probability that a part will become non-procurable at time t, assuming it was procurable in the interval $(0,t)$. Figure 4-12 shows $f(t)$; the corresponding hazard rate, $h(t)$, for linear regulators is shown in Figure 4-13. To determine $f(t)$, the data was fit with a 2-parameter Weibull,

$$f(t) = \frac{\beta}{\eta}\left(\frac{t}{\eta}\right)^{\beta-1} e^{-\left(\frac{t}{\eta}\right)^{\beta}} \tag{4-7}$$

where the parameters were estimated using MLE assuming right censoring and that the censoring mechanism is non-informative (the knowledge that the observation is censored does not convey any information except that the obsolescence dates of some parts within the data set lie beyond the censoring date, which is the analysis date [D_A] in our case).[5] In Figure 4-12, the uncensored distribution ignores the introduced parts that had not gone obsolete as of D_A. Obviously, the mode is shifted to the left (smaller procurement lifetimes) when the non-obsolete parts are ignored. In the case of linear regulators, the hazard rate shown in Figure 4-13 increases with time ($dh(t)/dt > 0$), indicating that the longer the procurement lifetime, the more likely the part is to go obsolete. In the case of the linear regulators, there are 347 obsolescence events out of a total of 847 introduced parts.

Figure 4-14 shows the quantity and fraction of non-obsolete linear regulators as a function of time. This plot shows that a large fraction of the parts

[5] The MLE parameter estimation was performed using the MatLAB Statistics package.

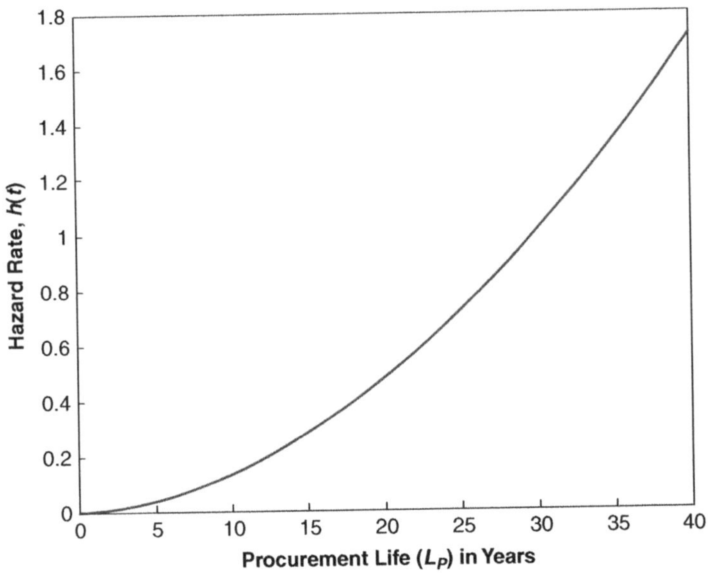

FIGURE 4-13 Hazard rate corresponding to the censored distribution of procurement lifetimes for linear regulators in Figure 4-11.

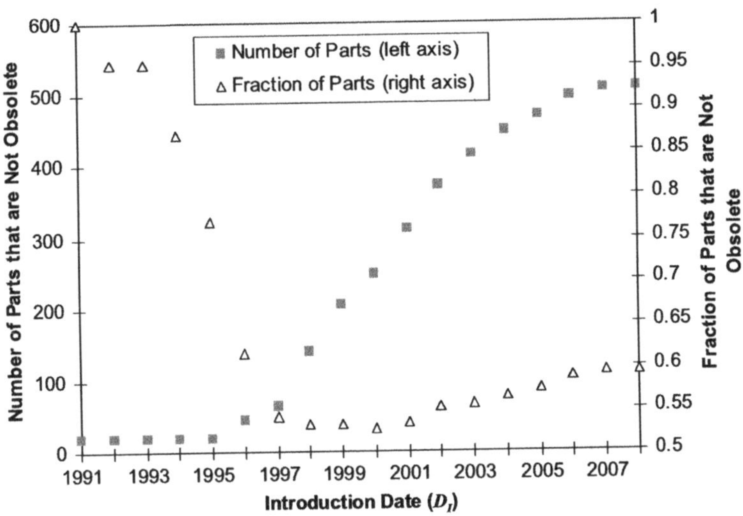

FIGURE 4-14 Quantity and fraction of linear regulators that were not obsolete as of 2008 as a function of time.

introduced in 1990–1996 had not gone obsolete by the date of assessment (but the total number introduced during this period was also relatively small). An alternative way to look at this is to perform the analysis described for determining the mean procurement lifetime on the data set as a function of time (Figure 4-15). In this case, to generate the mean procurement lifetime at a particular date (or before), we only consider the parts that had been introduced on or before that date in the analysis (although all observations were made in 2008). The mean procurement life is analogous to the mean time-to-failure (MTTF). The mean procurement life for a given Weibull distribution can be calculated using

$$\overline{L_p} = \eta \Gamma\left(1 + \frac{t}{\beta}\right) \tag{4-8}$$

where β and η are the Weibull parameters corresponding to the data fits up to D_I.

Figure 4-15 indicates the appropriate mean procurement lifetime to assume for parts with introduction dates at or before the indicated year. So a part introduced in 1998 or before has a mean procurement lifetime of 14 years (censored Weibull [2-P] in Figure 4-15). In order to determine the mean procurement lifetime for parts introduced in a particular year (rather than in or before a particular year), "slices" of the data must be used. In this case, to

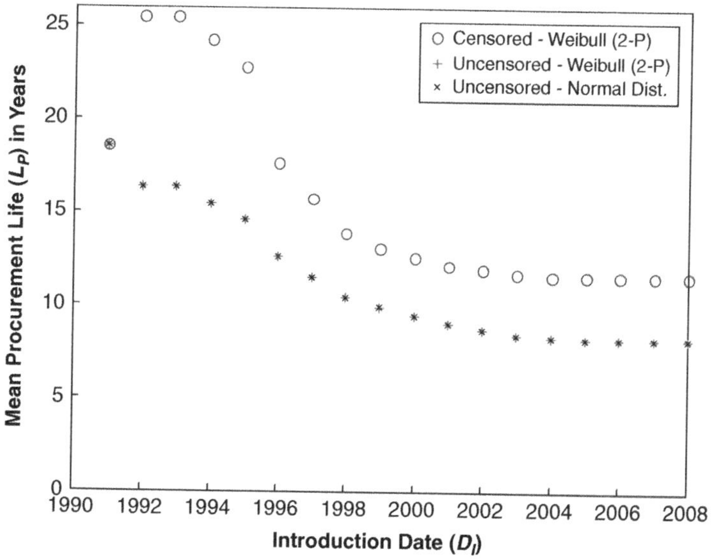

FIGURE 4-15 Mean procurement lifetime for linear regulators as a function of time (parts introduced on or before the date).

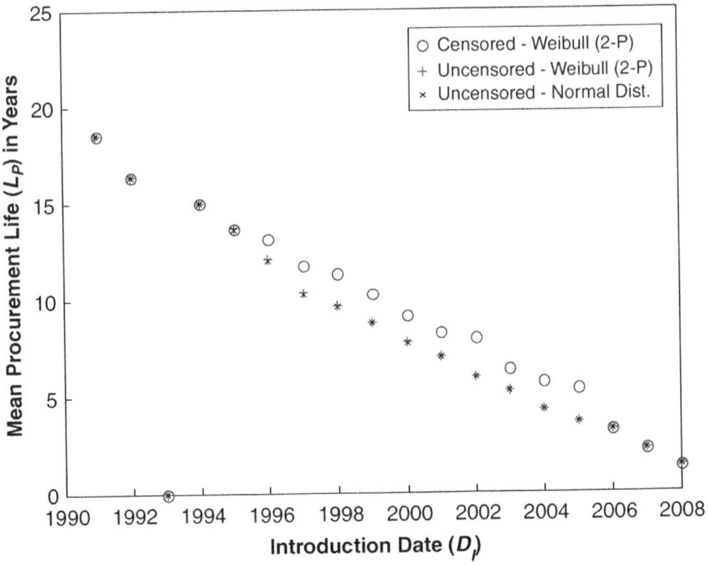

FIGURE 4-16 Mean procurement lifetime for linear regulators as a function of time (parts introduced on the date). Note: there were no parts introduced in 1993.

generate the mean procurement lifetime at a particular date, we only consider the parts that have been introduced within one-year periods in the analysis, and once again, all observations are made in 2008. Figure 4-16 shows the mean procurement lifetimes for one-year slices with and without right censoring, assuming that we are observing in 2008. For a part introduced in 1998, Figure 4-16 predicts that the mean procurement lifetime will be 11.5 years (smaller than the 14 years predicted by Figure 4-15). Figure 4-16 and the comparison of Figure 4-15 and Figure 4-16 indicate that older linear regulators (smaller D_I) have longer procurement lifetimes (L_P) than newer linear regulators.

Using the data from Figure 4-16 for 1990–2005 (excluding 1993, since no parts were introduced in 1993), the mean procurement life trend is given by

$$\overline{L_p} = 0.021089 D_I^2 - 85.2217 D_I + 86095.7 \tag{4-9}$$

The analysis in this section provides a useful estimation of the mean procurement lifetime for parts; however, the worst-case procurement lifetimes are also of interest to organizations performing proactive and strategic obsolescence management. The next section provides a more detailed interpretation of the procurement life versus introduction date profiles (see Figure 4-11) and discusses the generation of worst-case forecasts.

4.2.4 An Interpretation of Procurement Lifetime and Worst-Case Forecasts

Several distinct regions on the graph in Figure 4-11 can be identified. The upper bound of where specific data points (corresponding to specific observed obsolescence events) can lie in Figure 4-11 is given by $L_P = D_A - D_I$, where D_A is the analysis date, the date on which the analysis was performed (in Figure 4-11, D_A is 2008). The data set of known (observed) obsolescence events must always lie on or below this line. The region above this line represents parts that were introduced in the past but that are not obsolete yet. The region defined by $D_I > D_A$ could be populated by future part introductions. If the data set is complete (all parts that have gone obsolete up to D_A are included in the data set) then the region below the data and to the left of the $L_P = D_A - D_I$ line will be empty.

The top boundary of the procurement life versus introduction date plot changes with the addition of future obsolescence events, but the bottom boundary of the data set tends to stay constant over time as data is added. This is shown in Figure 4-17, where the 2008 analysis data includes all the obsolescence events that had occurred to date; the 2006 data includes all linear regulators that were introduced in 2006 or before and were obsolete in 2006 or before; and the projected 2010 analysis boundary is shown. In this case, the bottom boundary for parts introduced after 2005 is not known but can be projected based on the historical trend.

The top boundary of the data shown in Figure 4-11 will always have a slope of -1 because it is generated by a group of parts introduced on various dates

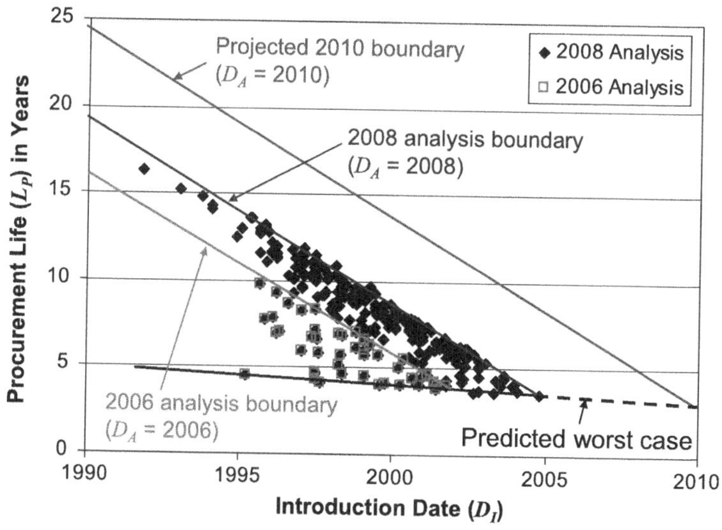

FIGURE 4-17 Linear regulator data separated by analysis date.

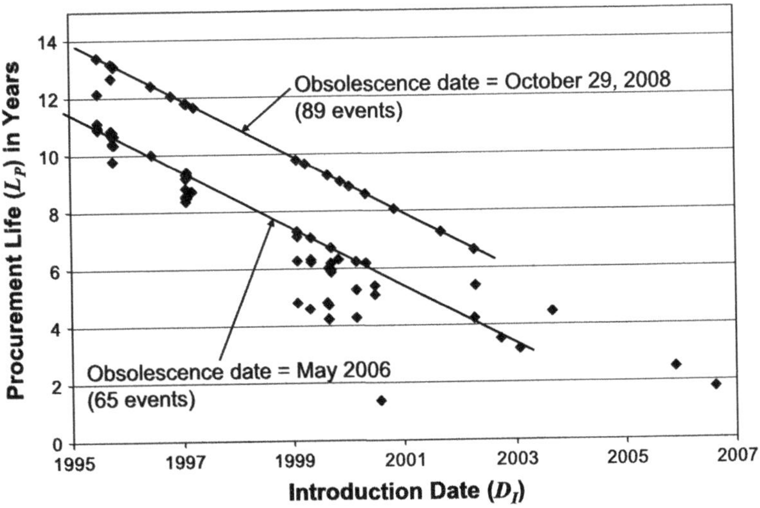

FIGURE 4-18 Integrated device technology logic devices (217 total obsolescence events).

that were all discontinued on or about the same date,[6] which is a common practice for electronic parts manufacturers (see Figure 4-18). A group of parts all having the same procurement life that are introduced on various dates would produce a horizontal line of data points on the plot. A vertical line of data points on the plot indicates that the parts all have the same introduction date and various discontinuance dates.[7]

The bottom boundary of the data shown in Figure 4-11 is the key to forecasting the worst-case procurement life. In cases where the "age" of the part (how long ago it was introduced) has no effect on the procurement life, the bottom boundary of the data set will be flat (slope = 0). If there is an age effect, the bottom boundary of the data set will have a non-zero slope. Parts with strong evolutionary parametric drivers have flat bottom boundaries. In the case of parts with strong evolutionary parametric drivers, the parametric driver usually causes the part to be discontinued before the procurement age has any impact. Figure 4-19 shows the procurement life versus introduction date for flash memory devices. The data set begins in late 1992 (earliest introduction dates). The bottom boundary of the data is flat. Flash memory obsolescence is driven by memory size, not by procurement age (see Sandborn et al., 2007a and Chapter 3). Alternatively, Figure 4-20 shows the procurement life versus introduction date for operational amplifiers (op amps). This characteristic

[6] The upper bound is defined by parts whose obsolescence date is D_A, and therefore the slope of the upper bound is given by $-(D_A - D_I)/(D_A - D_I) = -1$.

[7] In our experience, vertical lines of data points may be indicative of errors in the historical record. The most common error is that the introduction dates for the parts are the database record creation dates, not the actual part introduction dates.

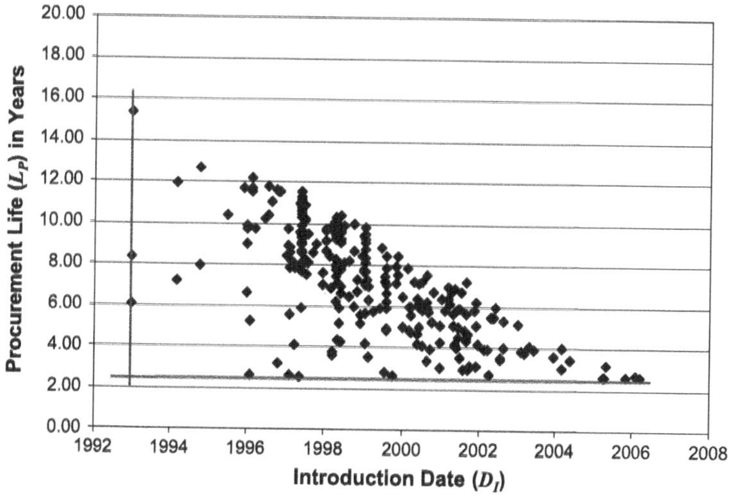

FIGURE 4-19 Flash memory devices (424 total obsolescence events from 38 manufacturers).

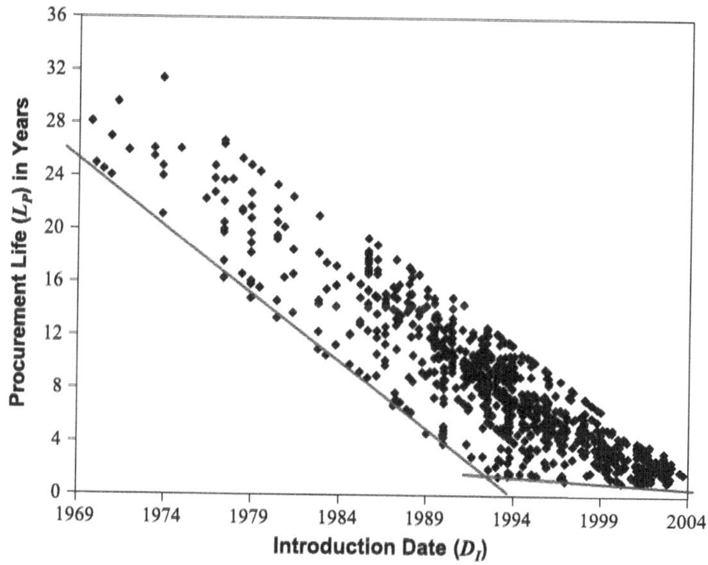

FIGURE 4-20 Operational amplifier devices (2400 total obsolescence events from 7 manufacturers).

clearly shows that for op amps introduced prior to 1993 there was a strong age effect; beginning in 1993, the age effect weakened but was still present.

For linear regulators shown in Figure 4-13, the bottom boundary (worst-case procurement lifetime) is given by:

$$L_P = -2.0947D_I + 4188.5 \text{ for } D_I \leq 1997.5 \tag{4-10}$$

$$L_P = -0.1014D_I + 206.77 \text{ for } D_I > 1997.5 \tag{4-11}$$

As indicated in Eqs. 4-10 and 4-11, the linear regulator procurement life trends show two distinct aging regimes. Other part types also often have multiple aging regimes, for example, operational amplifiers, as shown in Figure 4-20. The exact reason for the discontinuity is not known, however. Figure 4-14 shows that the year of the discontinuity for linear regulators (1997.5) is close to the minima in the fraction of parts that are not obsolete; and in both the linear regulator and operation amplifier cases, the year of the discontinuity in the procurement life trend also corresponds to a period of slow growth rate for the semiconductor industry (see Figure 4-21). Another possibility is that parts simply reached a minimum viable procurement life dictated by the high-volume products that demand the parts.

4.2.5 Part Type Specific Results

The analysis described in Sections 4.2.1 through 4.2.4 has been applied to a variety of electronic parts. Table 4-6 provides results for selected part types.

The methodology described in this section can also be applied to specific manufacturers, as shown in Figure 4-18 and Figure 4-22. Figure 4-22 shows the determination of the worst-case obsolescence forecast as a function of introduction year for linear regulators manufactured by National Semiconductor. The methodology can also be applied to key attributes of parts. Figure 4-23

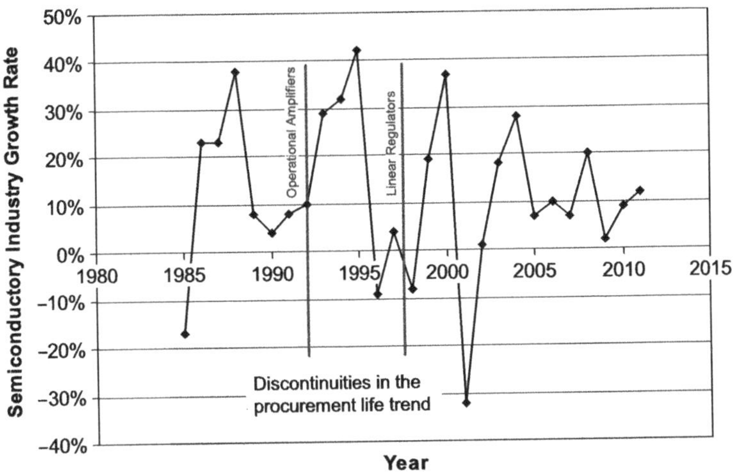

FIGURE 4-21 Correlation of discontinuities in the procurement life trends with the semiconductor industry growth rate (2007–2011 growth rates are forecasted) (McClean et al., 1999).

TABLE 4-6 Procurement Lifetimes for Various Electronic Part Types through 2008

	Censored				Uncensored					
Part Type	Mean (years)	β	η	*LKV*	Mean (years)	β	η	*LKV*	% of parts not obsolete	Total number of parts
Linear Regulators	11.63	2.842	13.06	−1205	8.245	3.470	9.168	−822	59.46%	509
Buffer & Line Drivers	38.39	2.021	43.33	−3042	9.821	4.008	10.83	−1293	91.08%	1279
Bus Transceivers	15.39	2.289	17.37	−1746	9.281	4.988	10.11	−984	57.26%	1057
Decoder & Demux	20.74	1.713	23.25	−914.9	9.298	4.505	10.19	−490	62.30%	565
Flip Flop	16.23	2.250	18.33	−1727	9.638	5.154	10.48	−950	58.27%	1052
Inverter Schmitt Trigger	18.13	1.830	20.40	−1125	8.575	4.061	9.453	−627	64.13%	750
Latch	14.99	2.289	16.92	−1391	9.186	4.952	10.01	−792	55.46%	818
Multiplexer	18.83	1.819	21.19	−844.0	8.835	4.096	9.734	−470	63.70%	552

(β and η refer to 2-parameter Weibull fits of the censored and uncensored PDFs. *LKV* is the negative log-likelihood function (larger negative values indicate a better fit.)

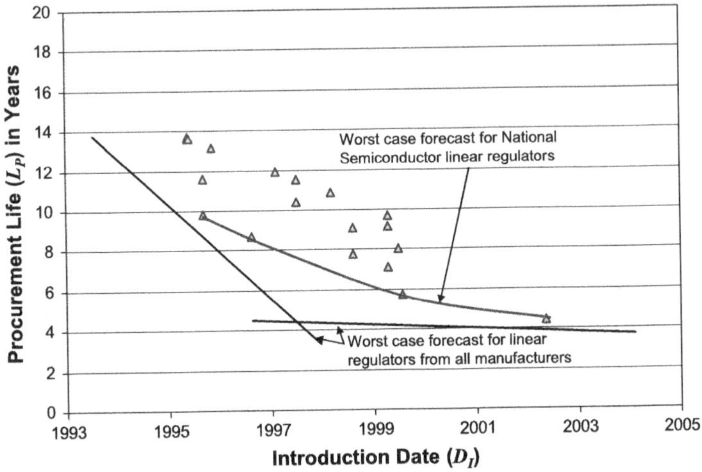

FIGURE 4-22 National Semiconductor linear regulator data.

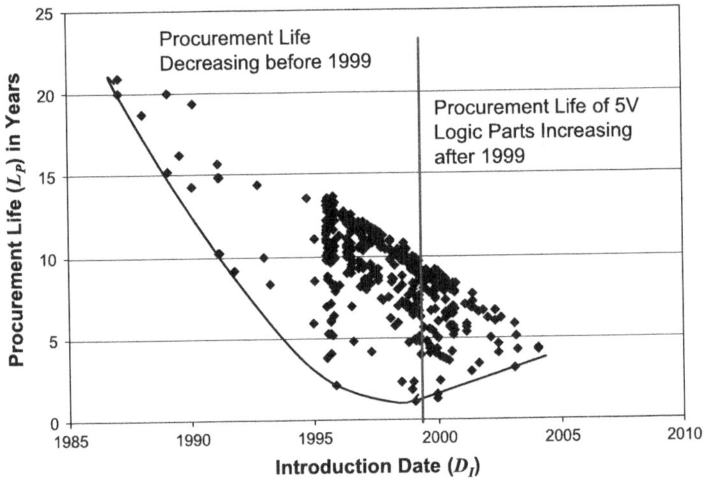

FIGURE 4-23 5V bias logic parts.

shows a procurement lifetime plot for 5V bias logic parts and is interesting because it clearly clarifies a decreasing procurement lifetime for parts introduced before 1999 (this is consistent with a broad trend from 5V bias parts to lower voltages, such as 3.3V and lower); however, the results also indicate that this trend may have reversed for parts introduced after 1999. This trend does not indicate that electronic parts are changing from lower bias levels back to 5V. Rather, it shows that 5V parts introduced after 1999 are being supported longer than their predecessors, which were phased out in favor of lower bias voltage versions. Manufacturers that introduce 5V parts are now targeting applications that are either not transitioning to lower voltage levels or whose conversion to lower voltage levels has been slow.

4.2.6 Discussion and Conclusions

In this section, we have presented a methodology for constructing algorithms that can be used to forecast the procurement lifetime, and thereby the obsolescence date, of technologies based on data mining of the historical record. Unlike the method described in Section 4.1 for forecasting DMSMS-type obsolescence, this method is applicable to technologies that have no clear evolutionary parametric driver. Results from the methodology applied to several different electronic part types have been included.

Long-range forecasting techniques generally involve methods of trend extrapolation. The worst-case procurement life trend for the linear regulator example developed in this section is shown in Figure 4-17 and quantified in Eqs. 4-10 and 4-11. Although trending the worst-case case procurement lives should be done using only the uncensored data (only the obsolete parts), trends in the mean procurement life must be performed using the censored data set (both obsolete and non-obsolete parts) from Figure 4-16 corresponding to Eq. 4-7.

It has been suggested that the "age" of electronic parts is not necessarily a factor in determining what becomes obsolete (Carbone, 2003). The age of a part can be interpreted two ways: either it represents how long ago the part was introduced ($D_A - D_I$), where D_A is the analysis date ($D_A - D_I$ is referred to as "design life" in Gravier and Swartz, 2009), or how long the part was procurable (L_P). The results in this section suggest that age is a factor in predicting the obsolescence of the part for parts that do not have strong evolutionarily parametric drivers. Gravier and Swartz (2009) also conclude that age is correlated to obsolescence by showing that the probability of no suppliers varies with design life. However, Gravier and Swartz did not distinguish between part types (except for military and non-military parts) or parts with or without strong evolutionary parametric drivers.

For the procurement lifetime forecasting algorithms developed using the methodology suggested in this section and Section 4.1 to be useful, one must assume that past trends are a valid predictor of the future. In some cases, particular technologies or parts may be displaced by some unforeseen new disruptive technology, thus accelerating the obsolescence of the existing parts faster than what the historical record would forecast. Alternatively, new applications may appear that extend or create demand for specific technologies or parts, also causing a change in the historical obsolescence patterns for the parts. Application of the proposed method depends on having access to sufficient historical data to support a statistical analysis; this is especially true when one wishes to refine the forecasts (with regard to particular vendors or particular part attributes).

4.3 NON-DATABASE OBSOLESCENCE FORECASTING METHODOLOGY

The forecasting methodologies described in Sections 4.1 and 4.2 depend on historical data for the electronic part. In some cases there is no historical data available. This is often the case when the items for which obsolescence needs to be forecasted are not electronic piece parts. The forecasting approach described in this section can be applied to subsystems (sometimes referred to as COTS assemblies) as well as to electronic piece parts.

The non-database obsolescence forecasting methodology described in this section does not primarily use statistical information to estimate product lifetimes; moreover, it aims at gathering planned and forecasted facts on the product from manufacturers directly. In order to manage large amounts of gathered obsolescence-forecasting-relevant data, the modification or introduction of an enterprise-resource-planning system (ERP system) is required.

4.3.1 Forecasting Process

The actual non-database approach contains three steps, as shown in Figure 4-24.

At a first glance this forecasting process seems to be easily manageable, but a lot of effort is required to get proper results. The necessary steps are described next.

4.3.2 Step 1: Identify Part/Technology Group

In this first step the part has to be identified according to Section 4.1.1.1. It is essential to know the part's main characteristics, such as function, manufacturer, and the preferred distributor. Due to effort reduction, some passive parts may be marked as irrelevant. For example, a common E96 SMT resistor with a commercial temperature range is manufactured by several sources. To determine if obsolescence forecasting for this part is necessary, its risk of obsolescence must be evaluated in the context of the project's circumstances. For instance, if a product's projected lifetime does not exceed five years total, one might abandon the forecast of a 100 Ohm SMT 0603 1 percent TK50 resistor (Tyco CPF0603F100RC1) (Tyco, 2010). This is only legitimate if the BOM does not specify the manufacturer. Labels and other drawing parts may also be

Step 1:	Identify Part
Step 2:	Obtain Forecasting Data
Step 3:	Estimated End-of-Life Date and Obsolescence Risk

FIGURE 4-24 Forecasting methodology Steps 1 to 3.

considered irrelevant. If there are any uncertainties, a forecast should be performed. There are some critical applications where no part is allowed to be valued as irrelevant, such as critical systems in aviation and medical applications.

4.3.3 Step 2: Obtain Forecasting Data

For this approach, the amount and quality of the input data are essential. Obtaining high-quality data takes a lot of time and effort, but it is necessary.

The most efficient way to gather forecasting-relevant data is to consult manufacturers directly as well as distributors. Contacting third parties such as component management databases or experts is another option, but their services may be costly. Consulting manufacturers and distributors is fairly simple. First, relevant information on the website should be collected, and, second, relevant information should be asked for directly (via telephone or e-mail).

To standardize this inquiry, the VMEbus International Trade Association released the open VITA 53.0 Standard for Commercial Technology Market Surveillance in 2009. This standard was defined by a group of experts from various companies (including Boeing, General Electric, Foxconn, Elma Bustronic, NSWC Crane) (VITA 53.0, 2009). This open standard contains obligatory questions for the manufacturer's inquiry form and can be individually adapted to a company's specific requirements.

Because this open standard is mainly focused on consumer products (such as laptops), it has limited applicability for inquires about electronic parts. Nevertheless, there are numerous similarities, and it can be used with some modifications as the basis for electronic parts as well. TQ Systems GmbH, for example, modified the open standard VITA 53.0 to meet its requirements for subassemblies and electronic components.

The difficulty with this way of gathering information is that some information (such as a part's share of manufacturer turnover) might be confidential or strategically motivated. Therefore, it is indispensable to get forecasting data from as many different sources as possible (see Table 4-7). For example, electronic manufacturing service companies aim at gaining forecasting-relevant data directly from manufacturers, distributors, and experienced application engineers. Only if the feedback from these three is consistent will the data be used in the forecasting process.

A basic life cycle forecast can only be performed if the low-grade data are provided. To improve the basic life cycle forecast's quality, more data are required. Medium-grade data allow an improved forecast result. The best case may only be achieved, if all 23 questions can be answered. The application of this data takes place in process Step 3.

Low-grade data contain the manufacturer market share and technology trends on the part's level (lower supply voltages or higher densities). For example, once a part is designed for automotive application, a long-term

TABLE 4-7 Required Data for Forecasting

Required Data No.	Fact	Low Grade	Medium Grade	High Grade
1.	Manufacturer market share	✓	✓	✓
2.	Technology trends	✓	✓	✓
3.	Part's typical applications	✓	✓	✓
4.	Part life cycle code according to EIA724	✓	✓	✓
5.	Part's replacement by same manufacturer	✓	✓	✓
6.	Part's replacement by different manufacturer	✓	✓	✓
7.	Number of authorized distributors of a part	✓	✓	✓
8.	Part's roadmap	✓	✓	✓
9.	Latest datasheet	✓	✓	✓
10.	Recommendations	✓	✓	✓
11.	Part's "long-term" program	✓	✓	✓
12.	Process technology roadmap	✓	✓	✓
13.	Manufacturer's focused market segments	✓	✓	✓
14.	Date of part's last revision change	✓	✓	✓
15.	Number of, dates of, and reasons for changes	✓	✓	✓
16.	Actual or estimated end of production date	✓	✓	✓
17.	Roadmap updates	×	✓	✓
18.	Number of part's revision changes	×	✓	✓
19.	Date of part's market introduction	×	✓	✓
20.	Part's share of distributor's turnover	×	×	✓
21.	Manufacturer's part sales data	×	×	✓
22.	Distributor's part sales data	×	×	✓
23.	Part's share of manufacturer's Turnover	×	×	✓

(Low-grade data enable a basic lifecycle forecast. Medium-grade data improve the basic life cycle forecast's quality. The best forecast quality may only be achieved if high-grade data are available.)

availability program (10–15 years of availability) might apply. Hence, it follows that the part's typical applications are useful facts. Furthermore, many manufacturers publish the current life cycle stages of their products. Identifying possible alternative parts from different manufacturers enables the comparison of gained data. Identifying the number of authorized distributors for the part results in multiple sources of information that can be compared. Another essential source for life cycle information is a product's roadmap issued by the manufacturer. Part datasheets may contain excellent data (date of initial release and revision time line). The documentation is not the only source of useful information; it might also be reasonable to ask the part's application engineers for information on long-term availability. Certain manufacturers supply "long-term" availability programs with guaranteed procurement life (Freescale, 2010b). Finally, the manufacturer's process technology roadmap and their focused future market segments are indicators of future decisions about the part's life.

To improve the methodology's forecasting quality, medium-grade information is required. This information contains the date of the part's last change as well as the total quantity and dates of previous changes. In order to reconstruct the change time line, it is useful to know the reasons for each change. For instance, a change in labeling information might not be relevant for forecasting the part's availability; but it is unlikely that an additional manufacturing or assembly site will be qualified if there is no intention to continue this part's production. If there is no information on changes available, analyzing the revision history of the latest datasheet might clarify some issues. The most important aspect is the date of the part's market introduction, which may correspond to the initial release of its datasheet.

Finally, high-grade data contain the part's share of the distributor's or manufacturer's turnover. Sales statistics (such as shipped units over past years) show the part's sales development process and directly apply to the life cycle model.

To gather information, market research organizations or component databases could be used at this stage as well. Another possible access route to information is to contact the manufacturer and distributor directly. Table 4-9 shows that it takes about two weeks to get 68.5 percent of all responses. Re-inquiring may be necessary to save time. The e-mail shown in Table 4-8 is a sample used by TQ Systems GmbH and could be sent to manufacturers and distributors.

Once manufacturers' feedback are registered and stored in the ERP system, the actual forecasting process can begin.

4.3.4 Step 3: Estimated EOL Date and Risk of Obsolescence

The data obtained in Step 2 needs to be checked for validity, since the manufacturers' and distributors' information might not be consistent. For example, the date of a part's market introduction might be described differently. This may be due to a simple misunderstanding, strategic motivation, or to there being several revision changes. If there are fundamental discrepancies, then another inquiry iteration should be considered. Finally, the comparison will demonstrate whether the information's quality is representative. Once this step is finished, all of the gathered data in Step 2 is validated.

To forecast the BOM of an industrial control unit containing about 250 parts (with 100 parts relevant for forecasting), the following results were registered by the Obsolescence Management Department within TQ Systems GmbH. Please note that the original BOM in Table 4-9 contains about 250 parts, which was then reduced (following Step 1) to 100 forecasting-relevant parts. The poor category in Table 4-9 refers to answers that contained no information or less information than required for low-grade data (see Table 4-7).

Basically, the projected end-of-life date provided will be used. If there is access to databases, it is reasonable (but optional) to check whether the given feedback correlates with life cycle or procurement lifetimes forecasts.

TABLE 4-8 Example Inquiry Form by TQ Systems GmbH

FROM:	Central Obsolescence Management Department
TO:	Manufacturer XYZ
SUBJECT:	Inquiry on future availability on product P/N: XXYYZZ

Dear Sir or Madam,

We would like to design in XXYYZZ. In order to support long-term availability for this project until YYYY, we would like to gather some information:

1. When was the market introduction?
2. How long is your company going to support this product?
3. Please give some typical application examples.
4. How would you describe the product's market development and the life cycle status according to EIA724?
5. Are any alternatives available?
6. Would you mind sending the latest update on the product's roadmap and datasheet?
7. Is there a special "non-obsolescence" policy for this product?
8. Are there any recommendations for products with long-term availability?
9. How many changes have occurred so far?
10. When was the last change?
11. Are sales increasing, constant, or decreasing?
12. What are the current lead times for mass production and samples?
13. Please explain the process for EOL and PCN issues.

If there are any questions, please do not hesitate to contact us. We are looking forward to your response by DD.MM.YYYY. Thank you for your kind support, which enables us to continue the design-in of your products in the future.

Yours sincerely,

NAME

TABLE 4-9 Quality, quantity, and response time of distributers' and manufacturers' feedback on forecasting data inquiries by the Obsolescence Management Department within TQ Systems GmbH

		Quality and Quantity				Response Time (in weeks)		
Inquiries to:	Qty	Poor	Low Grade	Medium Grade	High Grade	<1	<2	>2
Manufacturers	100	20	45	30	5	45	22	33
Distributors	100	39	48	13	0	18	52	30
Total	**200**	**59**	**93**	**43**	**5**	**63**	**74**	**63**
Percent of Total (%)		**29.5%**	**46.5%**	**21.5%**	**2.5%**	**31.5%**	**37%**	**31.5%**

Because 29.5 percent of the feedback in Table 4-9 did not contain sufficient information, statistical forecasting of life cycles or procurement lifetimes must be considered (see Sections 4.1 and 4.2). The well-recognized fact that most of the parts with poor feedback belong to material groups with no clear evolutional drivers simplifies the decision; for these parts, the procurement life forecasting model was constructed. As for the other 70.5 percent of the parts, the feedback data in Table 4-9 were low grade or better, and so that feedback can be applied to assessment.

4.3.5 ERP System Modification

Forecasting requires having the product's BOM with each part identifiable by a manufacturer's part number, its function, and product group. To cope with the large amount of data, an ERP system, such as SAP or Oracle, is required. This system must be capable of dealing with BOMs and saving information on the component level. To achieve high-quality forecasting results, the system must allow customer-specific modification.

Complex systems may contain many different printed circuit boards and therefore many different BOMs. In order to initiate Obsolescence Management, it is necessary to gather all relevant components in one BOM. To reach optimum efficiency, the forecasted data must be stored in the ERP system at the part level (such as in SAP MARA with transaction Code MM02, the Dynpros and MARA datafields must be created for this application). For economic reasons, the same parts are frequently used in many other BOMs as well. To store forecasting information in the central database prevents multiple forecasting for the same part (SAP, 2010).

Finally, with this single-level BOM the forecasting methodology can be applied to each single part. To reduce forecasting complexity, a software realization analog to SAP Workflow Agent enables multiple users to perform forecasts on different parts of one single-level BOM at the same time.

To get a reproducible description of parts' life cycle stages, the following standardized parameters are used:

1. Product life cycle (introduction, growth, maturity, decline, phase-out, obsolescence)
2. Product life cycle time (time from introduction to obsolescence)

Some parts do not require forecasting (see Section 4.3.2), so a forecasting relevance box should be implemented. It is also useful to store comments on the part level. The manufacturer may contribute additional information, such as alternatives or other references, which can be stored in this data field.

Historical data associated with obsolete parts of interest is required. This data can be generated by the ERP system, if introduction and discontinuance dates are available in the database. Another option is to contact the service provider. To begin, the data provided in the forecasting chapters can be used. It is reasonable to update this data from time-to-time.

All data gathered from manufacturers should be stored in a document management system linked closely to the ERP system in order to provide access to information within the company. This is important because more departments than just the Obsolescence Management Department are interested in the manufacturer's statements; for example, the purchasing and development departments also might be interested. With SAP ERP solutions, the application of a connected external document management tool is also possible.

4.3.6 Discussion and Conclusion

In comparison to the life cycle and procurement lifetime methodology, the non-database obsolescence forecasting methodology does not require a historical database. Instead of a statistical framework, the manufacturer's and distributor's statements are cross-checked, analyzed, and finally used for forecasting obsolescence issues. Because of the lack of historical data, this approach is more costly in terms of time, and there is a high dependency on information provided by manufacturers and distributers. This dependency can cause difficulties in response time (if the manufacturer requires two weeks to provide answers) and in the content of statements (the manufacturer refuses to forward information or if the information that is provided has been manipulated). On the other hand, this approach is appropriate to use when beginning to set up a historical database. Moreover, with this assessment, a supply chain network is being created, which can be used for further investigations. This methodology also provides a critical tool for project managers that enables them to forecast software obsolescence with some level of uncertainty (a confidence level) in order to mitigate the effects of software obsolescence.

Chapter 5

Case Study Hardware Forecasts and Trends

The forecasting methodologies described in Chapter 4 (Section 4.1.1) have been applied to different part types, including memory (DRAMs, SRAMs, EEPROMs, and flash memory), microprocessors, microcontrollers, logic, analog, and ASICs. For each part type, this chapter provides a brief overview of the part, discusses which forecasting methodology fits it best, demonstrates the application of the forecasting methodology, and discusses the forecasting results for that part type.

5.1 DYNAMIC RAMs (DRAMs)

Dynamic random access memory (DRAM) is memory that uses charge storage on a capacitor as an equivalent to binary data. The data can be accessed at any memory cell randomly rather than having to proceed sequentially from a starting cell. DRAM needs to have its storage cells refreshed (given a new electronic charge periodically) (Sharma, 1997). Compared with SRAM, DRAM has smaller structures and lower prices, but is slower. To compensate for the slower spend, synchronous dynamic random access memory (SDRAM) and double data rate (DDR) access types were introduced.

DRAMs are used as the main memory in most memory systems and have the highest volume production, covering about 16.2 percent of the total semiconductor market in 2006 (McClean et al., 2007). DRAM represents the least

Strategies to the Prediction, Mitigation and Management of Product Obsolescence, First Edition. Bjoern Bartels, Ulrich Ermel, Michael Pecht and Peter Sandborn.

expensive form of random access semiconductor memory. DRAM belongs to the volatile memory family because data is lost when power is removed.

5.1.1 Types of DRAMs

There are two categories of DRAM: asynchronous DRAM and synchronous DRAM, also referred to as SDRAM. Each category can again be further divided into a JEDEC standard conform memory and manufacturer-specific memory.

The main types of JEDEC standard conform asynchronous DRAM include fast page mode (FPM), extended data out (EDO), and burst mode (BM). The market share of FPM and EDO is no longer noteworthy, with only a few manufacturers still offering these in 2010 (Memphis, 2010a). BM never had a major market share. Manufacturer-specific memories include Windows RAM (WRAM) and Video RAM (VRAM). These types also never reached significant market shares. Because of SDRAM's faster access time, asynchronous DRAM lost its popularity.

The improved access performance of SDRAM is based upon its dual–bank architecture. Moreover, as the name implies, it is directly synchronized with the microprocessor's bus and therefore independent of extra buffering.

As for DRAM, the single data rate DRAM (SDR DRAM) is synchronized on rising or falling clock edge. Depending on the clock rate, there are two SDR DRAM: PC-100 and PC-133. Both types had less than a 2 percent market share in 2006 (McClean et al., 2007).

Compared to the SDR SDRAM, the market in 2006 was clearly dominated by DDR SDRAM (25 percent market share) and DDR2 SDRAM (63 percent market share). DDR2's successor is DDR3, which gained market share in 2010 (for example, Apple's Macbook Air contains a 2GB DDR3 SDRAM [Apple, 2010]). DDR (double data rate) refers to the fact that both the rising and falling edge of the clock pulse carry data, thus doubling the data rate.

There is a clear trend in increasing speed and minimizing structures, but reducing supply voltages has become increasingly important. To make their memories attractive for mobile devices, a variety of manufacturers introduced SDRAMs with reduced power consumption. Since there was no clear standard, these memories show manufacturer-specific characteristics. Other manufacturer- specific memories include embedded DRAM, customized DRAM, cache DRAM (CDRAM), enhanced DRAM (ESDRAM), virtual channel DRAM (VCDRAM), reduced latency DRAM (RLDRAM), mobile RAM, COSMO-RAM, low-power DRAM (LPDRAM), pseudo static RAM (PSRAM), protocol-based DRAM, synclink DRAM (SLDRAM), direct rambus DRAM (DRDRAM), and XDR DRAM.

Some vendors and their DRAM product portfolios are listed in Table 5-1.

5.1.2 Market and Technology Trends

Due to the application of DRAM as the main memory in computing systems, operating systems are the biggest wildcard for the DRAM market. The connection

TABLE 5-1 Major DRAM Vendors and DRAM Types

Vendor	DRAM Architectures Offered
Samsung	DDR3, DDR2, DDR, SDR, LPDDR2, mobile SDR, mobile DDR, XDR, GDDR1, GDDR2, GDDR3, GDDR4
Micron	DDR3, DDR2, DDR, SDR, RLDRAM, mobile LPDRAM, PSRAM
ChiPlus	SDR
AMIC	SDR, mobile DRAM, low-power EDO, FPM, DDR1
Eplida	DDR2, DDR3, mobile DDR2 RAM, GDDR3, LPSDR, XDR, LPDDR, LPDDR2
Eorex	mobile SDRAM, mobile DDR, SDR, DDR, DDR2
Etron	DDR2, DDR, SDR, LP DDR, LPSDR, mobile DDR, mobile SDR, VLP PSRAM
Fidelix	PSRAM, LP SDR, LP DDR
Hynix	DDR3, DDR2, DDR, SDR, GDDR5, GDDR4, GDDR3, mobile SDR, mobile DDR, mobile DDR2, PSRAM
Promos	DDR2, DDR, SDR, mobile DRAM
Winbond	PSRAM, LP SDR, LP DDR, SDR, DDR, DDR2, DDR3, GDDR3, GDDR5
Nanya	DDR3, DDR2, DDR, SDR, mobile LP SDR, mobile LP DDR, mobile LP DDR2
Powerchip	DDR3, DDR2, DDR, SDR

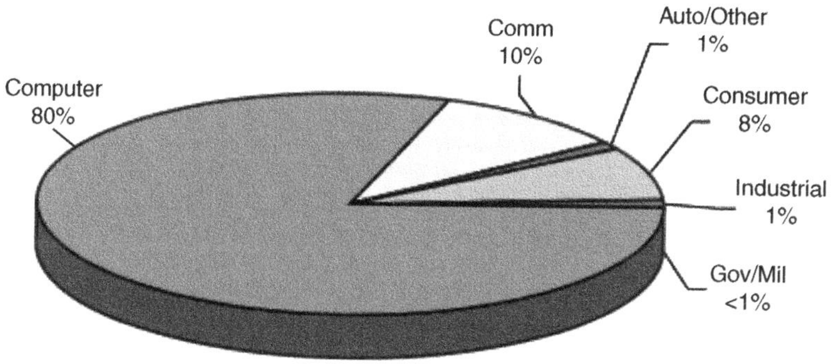

FIGURE 5-1 The 2006 DRAM market distribution by application (McClean et al., 2007). *Source:* WSTS, IC Insights

to the computing market (see Figure 5-1) is the price, whereas technology and demand are the driver. Moreover, because the applications of DRAMs are so heavily driven by consumer electronics (like the Nintendo Wii), most DRAM types are purchasable for relatively short periods of time.

In general there are two technology trends that cause DRAMs to evolve: the data exchange rate is increasing (see Figure 5-2 and Figure 5-3) and the power consumption is decreasing (see Figure 5-4).

Depending on the memory's application, one or more of these trends apply. For example, the DRAM for PC main memory is primarily driven by density

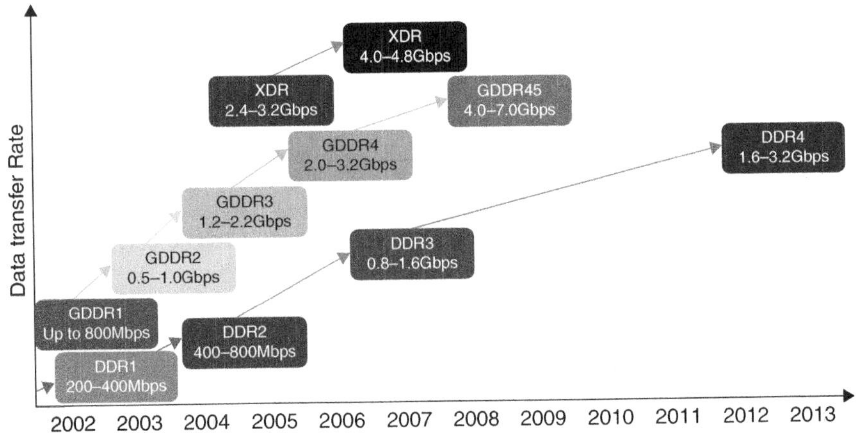

FIGURE 5-2 Data transfer rates for DRAM types by Samsung (Samsung, 2010a).

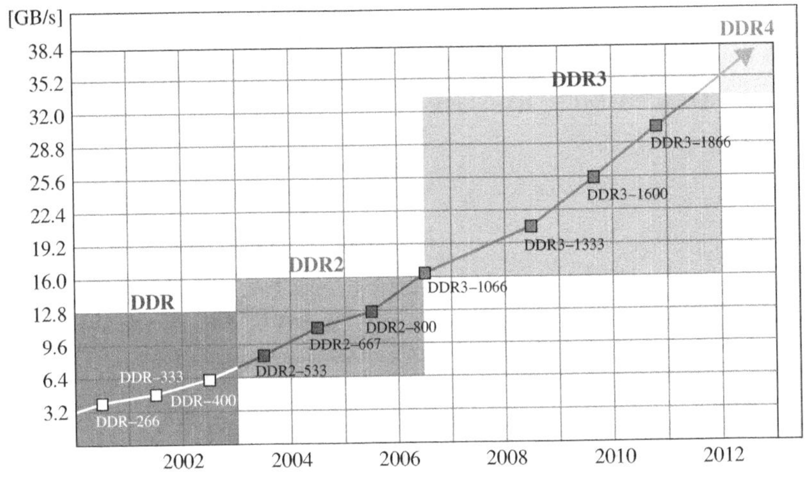

FIGURE 5-3 Evolution of DDR main memory data transfer rates for computing (Samsung, 2010a).
Source: Samsung

and speed, the memory of mobile devices by power consumption and density. The DRAM for graphic processing is dominated by data transfer rates.

In 2006, the total DRAM market grew 33 percent and the average selling price (ASP) jumped 18 percent (McClean et al., 2007). The introduction of new operating systems (Microsoft Windows 7) and new hardware (Apple's Macbook Air) are strongly correlated to the DRAM market growth rate. The increased ASP is due to reduced manufacturing capabilities and constant demand. DRAM manufacturing lines can be modified for flash memory production, which requires multi-million investments. With less investment, old DRAM production lines can be used to build CMOS sensors (Hynix, 2010).

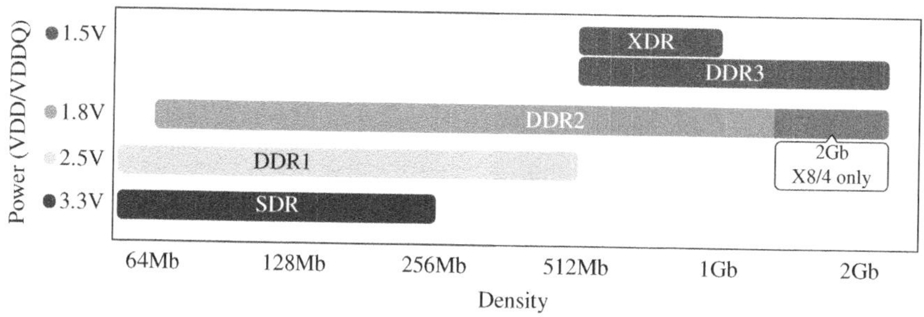

FIGURE 5-4 Supply voltages for DRAM types and densities by Samsung (Samsung, 2010a).

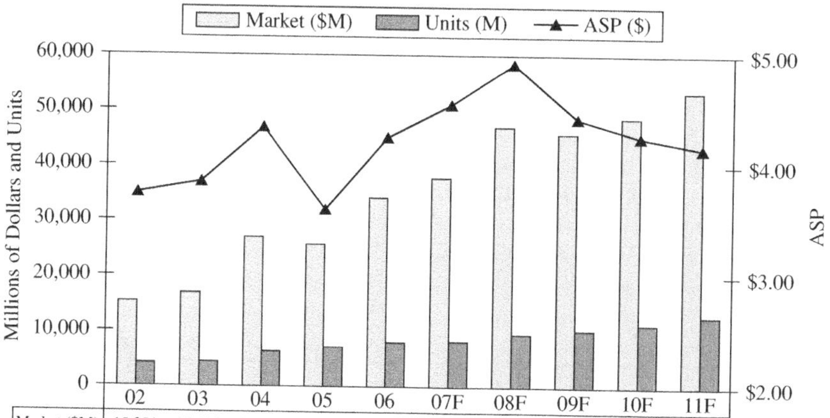

	02	03	04	05	06	07F	08F	09F	10F	11F
Market ($M)	15,254	16,688	26,849	25,596	34,050	37,775	47,000	45,600	48,600	53,300
Units (M)	4,066	4,331	6,182	7,105	8,000	8,300	9,550	10,300	11,400	12,800
ASP ($)	3.75	3.85	4.34	3.60	4.26	4.55	4.92	4.43	4.26	4.16

FIGURE 5-5 The DRAM market (McClean et al., 2007).
Source: WSTS, IC Insights

Nevertheless, as seen in Figure 5-5, the DRAM market is evolving strongly.

To persist in this dynamic market, the manufacturers must cope with the increasing cost of production and equipment as well as declining prices per bit.

5.1.3 Application of the Forecasting Methodology

Because a clear evolutional driver (in this case, memory density) dominates DRAM evolution, the Basic Life Cycle Forecasting Methodology (see Section 4.1.1) will be used. Of course, the Advanced Life Cycle Method could also be applied (see Section 4.1.2), but this methodology will be applied later to flash memories.

To forecast both the years to obsolescence and the life cycle stage, a model based on sales volume is used. A primary attribute determines the curve-fit to the sales data to the model, and modifying attributes translate the obsolescence predictions, depending on their degree of influence over the primary attribute.

TABLE 5-2 Sample Calculation for the Modified Zone of Obsolescence of a 16Mb, 5V, SOP DRAM, Date of Assessment $p = 1999$

Attribute	μ (μ_j for modifying attributes)	σ(σ_j for modifying attributes)	$(\mu + 2.5\sigma - p, \mu + 3.5\sigma - p)$	Comparision of limits	Modified zone of obsolescence	Comments
Main: Megabit size (16Mb)	1998.2	1.6	(3.2, 4.8)			
Modifying: Package style (SOP)	1999.7	5.2	(13.7, 18.9)	18.9 > 4.8	(min [13.7,3.2], 4.8) = (3.2, 4.8)	No effect
Modifying: Voltage (5V)	1990.4	4.3	(2.1, 6.4)	6.4 > 4.8	(min [2.1,3.2], 4.8) = (2.1, 4.8)	Modified

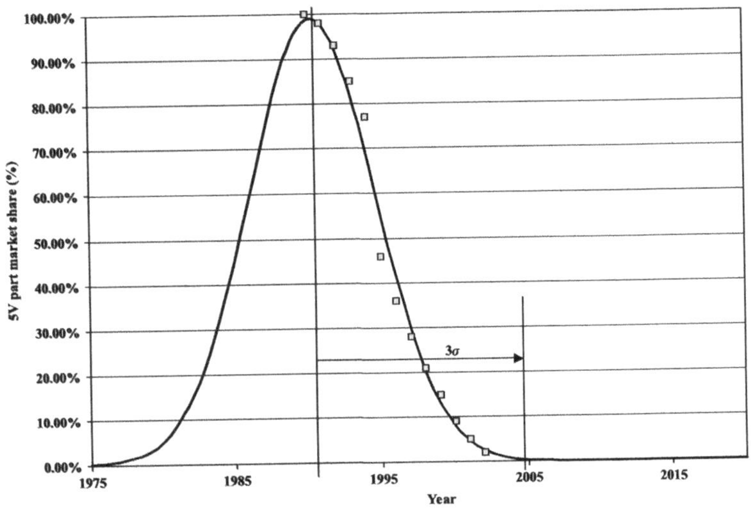

FIGURE 5-6 Life cycle curve of any 5V part.

In the case of DRAMs, the primary attribute is the DRAM memory density, and the dominant modifying attributes are package style and nominal supply voltage. The forecasting methodology described in Chapter 4 delivers the results shown in Table 5-2 and are printed in Figure 5-6 and Figure 5-7.

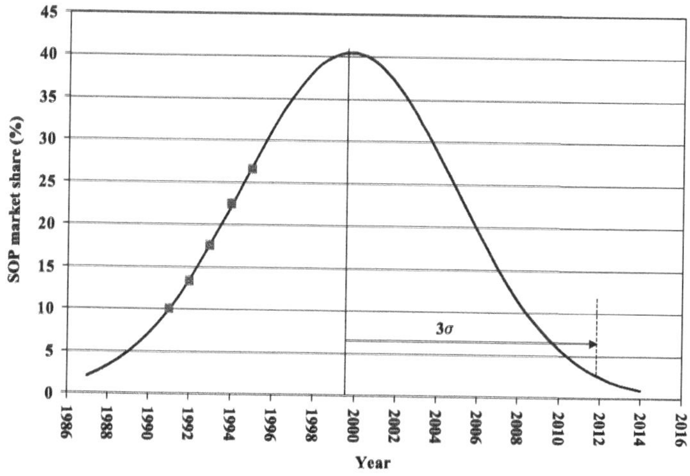

FIGURE 5-7 Life cycle curve of a small outline package (SOP).

5.1.4 Discussion of DRAM Forecasts

Chapter 4 discusses DRAM forecasts as a function of memory size based on sales data through 2006. DRAM shipments occur in cycles that progressively shorten with increasing technology generations. The peak sales of each technology generation are observed to be higher due to the rapid growth and large market share in the PC industry.

In 1999, the life cycle forecasting methodology projected that the obsolescence of the 16Mb 5V SOP DRAM would occur between 2001 and the end of 2003. However, there were still units shipped in 2006. Of course, the number of units shipped in 2006 decreased 88.1 percent relative to 1999, but still there were about 250 million units shipped. At first glance, the forecasting model did not match reality. There was even a slight increase in units shipped between 2003 and 2004, which in 1999 was forecasted to be the zone of obsolescence (see Figure 5-6).

The question to be resolved is, what caused this discrepancy of two years between the forecast and the actual shipment's drop to less than 4 percent? First, forecasting will never predict the future exactly. This forecasting approach uses historical data and projects trends into the future. The methodology assumes that the future will follow the past trend exactly, which in reality it will not. Moreover, the dynamics of IC markets are highly volatile compared with other markets.

But why is this obvious difference almost two entire years? To clarify, let's consider the manufacturer's strategy. Next-generation DRAMs become popular only when their price reaches the "crossover" point, which is where the cost per bit of the new generation drops below the cost of the cost per bit of the current generation of memory chips. The 64Mb DRAM, which was introduced in the first quarter of 1997, reached the crossover point around the second quarter of 1999. The delay in reaching the crossover point was because Micron, the second biggest supplier of DRAMs, continued to produce 16Mb DRAMs at costs that were cheaper than the 64Mb DRAM, "pushing out" the crossover point. In certain applications, the total costs of memory are essential. Although the costs per bit might already have reached the crossover point, the memory in total is still more expensive than its successor. Therefore, the smaller memory will experience a longer market presence.

There have been applications in consumer electronics that simply do not require more memory than the 16Mb. Moreover, smaller packages, reduced voltages, and its low price continued to make the 16Mb still interesting after its originally forecasted obsolescence date.

Although the life cycle forecasting methodology forecasted the obsolescence date much earlier, this approach is still highly relevant, because it represents the majority of memories. And it is true that after 2001 the sales of 16Mb DRAM represented less than 45 percent of the total DRAM shipment data. In summary, the life cycle forecasting approach is not faulty, but one should keep in mind that the results should be referred to as trend prognosis for the majority of memories. Furthermore, in this case, the window of obsolescence defined in the basic method is not quite right for DRAMs.

The advanced method in Section 4.1.2 corrects some of the issues discussed here by deriving the window of obsolescence from data mining rather than arbitrarily setting it, as in the basic method. The basic method is necessary but not sufficient to forecast the obsolescence of these parts.

5.2 STATIC RANDOM ACCESS MEMORIES (SRAMs)

Static random access memory (SRAM) retains data bits in its memory as long as power is supplied. Once the supply voltage is switched off, all data will be lost, which means that SRAM, like DRAM, belongs to the volatile memories family (non-volatile memories include flash and PROMs, which will be discussed in Section 5.3.1). Unlike DRAMs, which store bits in cells consisting of a capacitor and a transistor, SRAMs do not have to be periodically refreshed. SRAMs store data in a flip-flop consisting of logic transistors and are ready for another read cycle as soon as the first is complete. Unlike SRAMs, DRAMs have to write data back into the cell because data is lost after a read cycle. Hence, SRAMs provide faster access to data than DRAMs. Although SRAMs are faster than DRAMs, the memory capacity of SRAMs is approximately 25 percent of DRAMs for the same chip size and process technology. This is because SRAM transistor structures require more space than capacitors. The cost of memory is largely determined by chip size, so SRAMs cost almost four times more per bit than DRAMs (Garrett, 2000). Figure 5-8 shows that DRAM memory cells only require one transistor and one capacitor, whereas a standard SRAM cell requires six transistors. As for process technology and area requirements on wafers, the DRAM cell is predominant.

Even though SRAMs have faster access times, SRAMs are not a major player in the memory market today. The market is dominated by DRAM memories, because

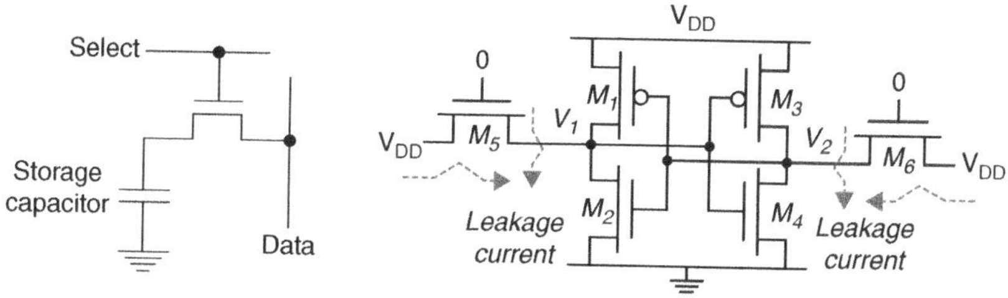

FIGURE 5-8 Comparison of DRAM (left) and SRAM (right) structures for a single memory cell (Qin, 2010) (Usenix, 2010).

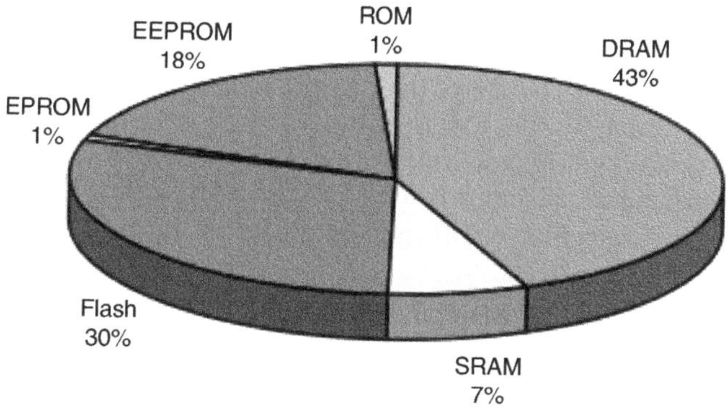

FIGURE 5-9 Worldwide 2006 memory units shipments (McClean et al., 2007).
Source: WSTS, IC Insights

of their higher densities. As shown in Figure 5-9, in 2006 SRAM only accounted for 7 percent of worldwide memory shipments (McClean et al., 2007). SRAM in combination with DRAM or flash is especially popular in the mobile phone sector, where size, weight, power consumption, and form factors are essential.

5.2.1 Types of SRAMs

Like DRAMs, the major types of SRAMs are synchronous and asynchronous. Synchronous SRAMs synchronize their memory speed with the microprocessor speed. Alternatively, asynchronous SRAMs are not synchronized with the microprocessor's clock frequency.

There are two ways to classify SRAMs. The first is to classify them by access time and divide them into fast and slow (Memphis, 2010b). Fast SRAMS have less than 10 ns access time (Memphis, 2010b). Another way to classify SRAM is by their technology. There are low-power SDRAM, burst SDRAM, pipelined burst SDRAM, no-turnaround (or zero-bus-turnaround or zero bus latency or no-wait SRAMs), double data rate (DDR SRAM), and quad data rate SRAM (QDR SDRAM). See Table 5-3.

TABLE 5-3 2010 Typical Access Times for Various Fast SRAM Densities (MSC, 2010)

SRAM Density	Access Time (ns)
16Kb	20
64Kb	15–20
2Mb	8–15
8Mb	8–20
16Mb	8–20
64Mb	15

The principal volume applications for slow SRAM are cellular phones and applications requiring the retention of data during the loss of main power. Slow SRAMs have low power consumption, so data can be stored by using a backup battery in the system. Therefore, the backup power source life is maximized. Another application of SRAM is advanced network equipment (including high-speed network switches, routers, and other telecommunication equipment). For instance, Samsung's K7R643684M SRAM (72Mb QDR II SRAM with 1.8V and 165pin FBGA) offers a 0.45 ns access time (Samsung, 2009).

5.2.2 The SRAM Market

The 2006 SRAM market shows clear dominance by three application segments. First, many workstations still require discrete cache memory, which leads to a demand for high-speed and high-density SRAM. In 2006, fast 400–500 MHz SRAM with densities greater than 1 Mb were the rule (McClean et al., 2007). Moreover, Internet and communication infrastructure equipment demands high-speed SRAMs (Samsung, 2009). Finally, mobile phones and handsets contribute to a constant demand for SRAM. Especially compared to DRAM, low access times are considered to be the unique advantage of SRAM.

According to Figure 5-10, sales and shipments stayed constant between 2004 and 2011. The last increase in shipments was between 1998 and 2004. After this period, the increase turned into a constant level at about 1,250 million units shipped. Besides density improvements, the access time also decreased significantly. In 2000, a 4Mb SRAM had a 15 ns access time. In 2010, a 4Mb SRAM had 8 ns access time. The QDR II technology already offers high-speed access down to 0.45 ns.

The 2006 SRAM market was dominated by PSRAM and fast SRAM (see Figure 5-11). The PSRAM (see Section 5.1.1) actually belongs to synchronized DRAM, but due to its SRAM-like interface it is considered to be an SRAM. The advantages of this memory type are its low power consumption, small footprint, and low price. As far as speed, it is not seen as the differentiator. The major demand for PSRAM is from the mobile phone market. Due to this market's high amount of shipments annually, the PSRAM dominated the SRAM market in 2006 with 52 percent (McClean et al., 2007). About

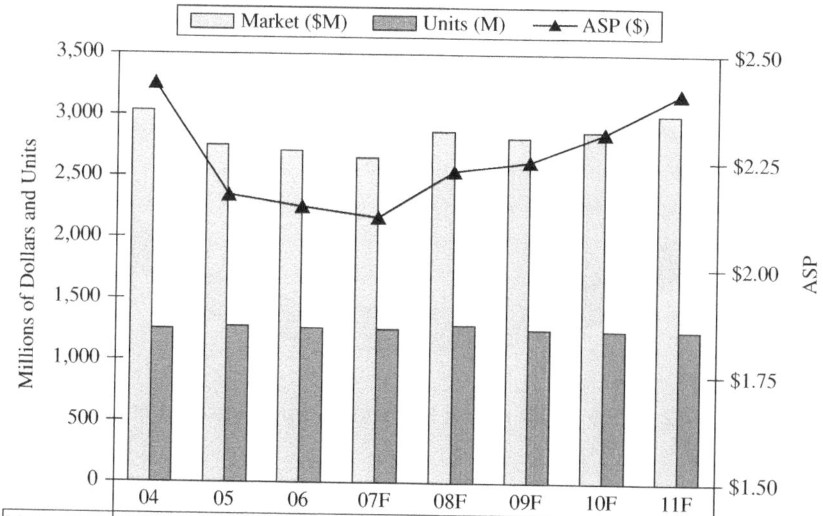

	04	05	06	07F	08F	09F	10F	11F
Market ($M)	3,032	2,754	2,710	2,656	2,868	2,811	2,867	3,010
Units (M)	1,246	1,269	1,265	1,252	1,288	1,250	1,238	1,250
ASP ($)	2.43	2.17	2.14	2.12	2.23	2.25	2.32	2.41

FIGURE 5-10 Historic and forecasted SRAM sales trends (McClean et al., 2007).
Source: WSTS, IC Insights

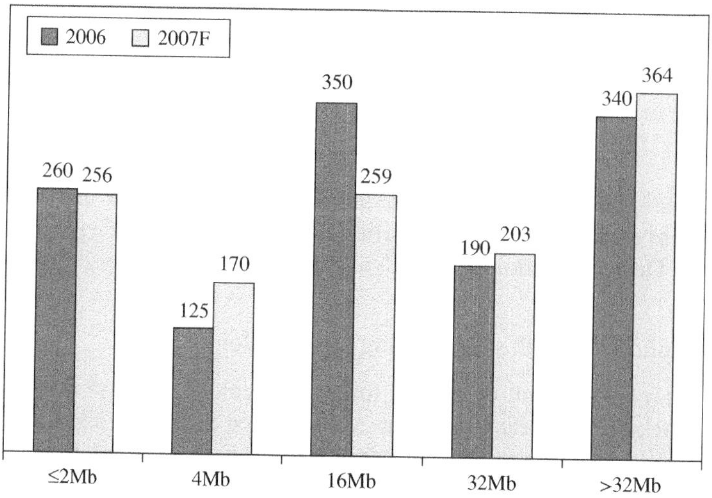

FIGURE 5-11 2006 SRAM shipments (in millions) by density (McClean et al., 2007).
Source: WSTS, IC Insights

11 percent of the SRAM sold in 2006 belong to the high-speed market. The definition of high speed (less than 30 ns access time) no longer applies in 2010. In 2010, all SRAM with less access time than 10 ns were considered to be high speed (Memphis, 2010b). See Figure 5-12.

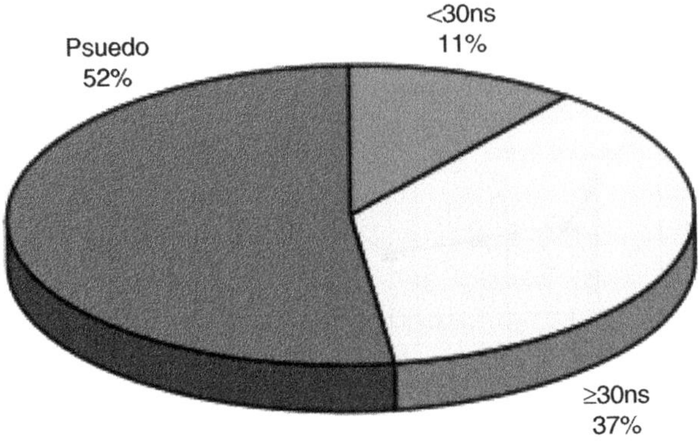

FIGURE 5-12 2006 SRAM market by memory types (McClean et al., 2007).
Source: WSTS, IC Insights

TABLE 5-4 Leading SRAM Manufacturers in 2010 (including PSRAM)

Alliance	ISSI
AMIC	Renesas
Chiplus	Samsung
Cypress	BSI
Etron	Microchip
Fidelix	IDT
Hynix	Micron
Lyontek	

In Table 5-4 all major SRAM manufacturers in 2010 are listed. The development of the next QDR generation SRAM is driven by Cypress, Renesas, Integrated Device Technology, NEC, and Samsung (Samsung, 2010b).

5.2.3 Application of the Forecasting Methodology

Since a clear evolutional driver (in this case memory density) dominates the SRAM market, the Basic Life Cycle Method (see Section 4.1.1) can be used. Like the DRAM forecast, the Advanced Life Cycle Method (referred to in Section 4.1.2) could be applied as well. Nevertheless, for DRAM and SRAM forecasting we focus on the Basic Life Cycle Method. Flash memory, discussed later, conforms to the Advanced Life Cycle Method.

In the case of SRAMs, the primary attribute is the SRAM memory size or density, and the dominant modifying attributes are package style and nominal supply voltage. Data from IC Insights (McClean et al., 1999) and McClean (2007) were used to develop the algorithm to forecast the life cycle stage and years to obsolescence of an SRAM part. See Figure 5-13 and Table 5-5.

The type of SRAM may modify the algorithm, or life cycle curve, of SRAMs based on memory size. The major SRAM types in 2000—non-cache SRAM, SRAM with cache, and sync-burst SRAM with cache—have been used in 16-bit, 32-bit, and 64-bit CPUs, respectively, to improve PC performance. Data from IC Insights has been used to model the distribution parameters, mean (μ) and standard deviation (σ), for the SRAM types, which are given in Table 5-6.

Finally, the forecasting methodology described in Chapter 4 delivers the results in Table 5-7.

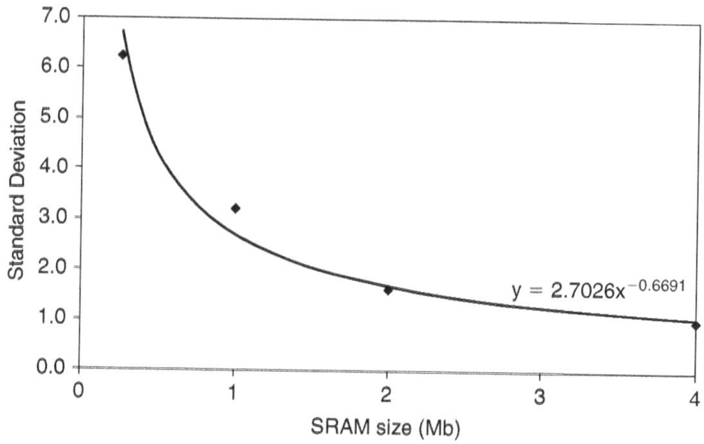

FIGURE 5-13 Standard deviation versus SRAM memory density.

TABLE 5-5 Mean (μ) and Standard Deviation (σ) for SRAMs by Bit-Size (Primary Attribute), p = Date of Assessment

Memory Size	Mean μ	Standard Deviation σ	$\mu + 2.5\sigma - p$	$\mu + 3.5\sigma - p$
64Kb	1981.3	7.3	0.55	7.85
256Kb	1996.3	2.6	3.5	6.1
1Mb	1999.6	3.1	7.75	10.85
> 1Mb	2001.2	2.2	7.7	9.9

TABLE 5-6 Mean (μ) and Standard Deviation (σ) for Secondary Attributes for SRAMs (p = Date of Assessment)

SRAM Type	Mean μ	Standard Deviation σ	$\mu + 2.5\sigma - p$	$\mu + 3.5\sigma - p$
Non-cache SRAM	1990.3	3	−1.2	1.8
SRAM with cache	1994.8	2.2	1.3	3.5
Sync-burst SRAM with cache	2000	2.4	7.0	9.4

(Continued)

TABLE 5-6 Continued

SRAM Type	Mean μ	Standard Deviation σ	$\mu + 2.5\sigma - p$	$\mu + 3.5\sigma - p$
Package style				
DIP	1975.7	8.6	−1.8	6.7
SOP	1999.7	5.2	13.7	18.9
CC	1995.3	2.7	3.0	5.8
PGA	1996.7	4.5	9.0	13.5
QFP	1997.3	3.6	7.3	10.9
MCM	1997.5	3.6	7.5	11.1
Other package styles[1]	1998.6	4.7	11.3	16.1
Voltage				
5V	1990.4	4.3	2.1	6.3
3–5V	1996.5	2.9	4.7	7.6
3–3.5V	1999.7	3.3	8.8	12.0
< 3V	2003.6	3.6	13.6	17.1

TABLE 5-7 Forecasting Results of Sync-Burst 1Mb SRAM with 3.3V in QFP Package Performed

Attribute	μ (μ_j for modifying attributes)	$\sigma(\sigma_j$ for modifying attributes)	$(\mu + 2.5\sigma - p, \mu + 3.5\sigma - p)$	Comparision of limits	Modified zone of obsolescence	Comments
Main: Megabit size (1Mb)	2000	2.4	(7.0,9.4)			
Modifying: Package style (QFP)	1997.3	3.6	(6.3,9.9)		(min [7.0,6.3], min [9.4,9.9]) = (6.3, 9.4)	Modified
Modifying: Voltage (3.3V)	1999.7	3.3	(7.95, 11.25)		(min [6.3,7.95], min [9.4,11.25]) = (6.3, 9.4)	No effect

[1]Includes chip-scale packages (CSPs), ball-grid-array packages (BGAs), and micro-ball-grid-array packages (μBGAs).

5.2.4 Discussion of SRAM Forecasts

The SRAM life cycle curves by memory size are shown in Figure 5-14 and by SRAM type are shown in Figure 5-15. Figure 5-14 depicts that gradual demise of lower memory densities (≤64Kb) and the appearance of higher memory densities with time. Figure 5-15 shows the demise of non-cache SRAM and the growth of cache SRAM and sync-burst with cache SRAM.

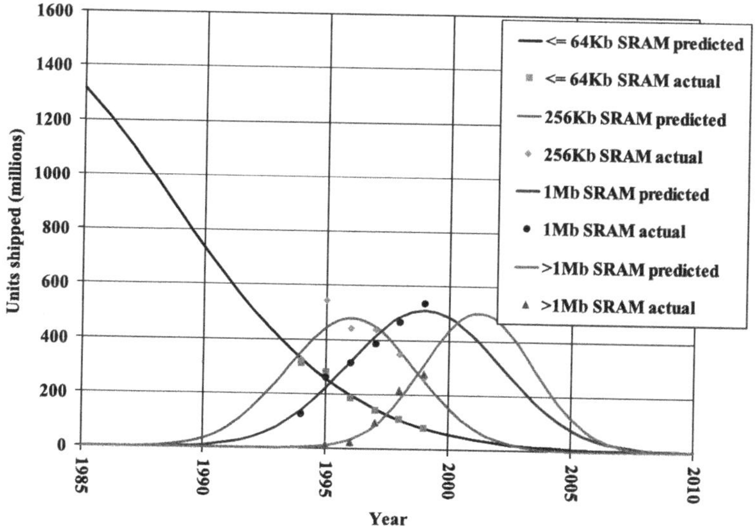

FIGURE 5-14 SRAM Life cycle curves by memory size.

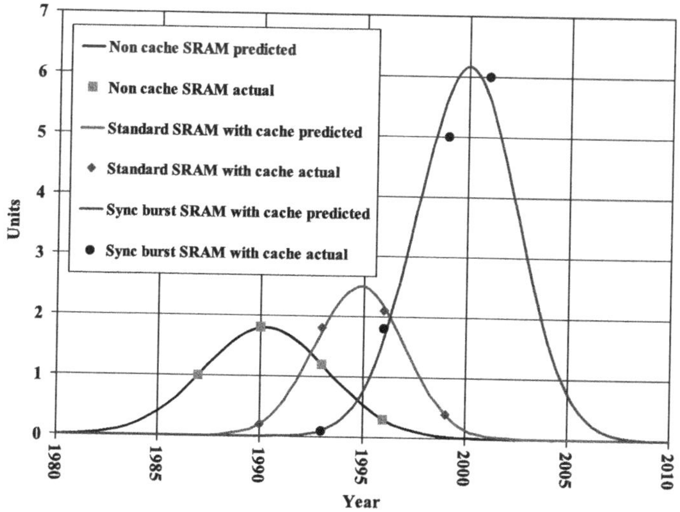

FIGURE 5-15 Life cycle curves of SRAM by SRAM type.

SRAM vendors face significant competition from emerging technologies. Every few years, DRAM technology advances in speed, and with its inherent cost advantages, DRAM displaces the slower performance spectrum of SRAM products. SRAM designers concentrate on advancing the memory speed over DRAM and increasing memory density. For example, in 2010 Samsung Semiconductors offered a 4Gb DDR3 DRAM with 1.875 ns access time (K4B4G0846A) and a 72Mb SRAM with 0.45 ns access time (K7R641884M) (Samsung, 2010c) (Samsung, 2010b).

SRAM modules are available in densities up to 72Mb (MSC, 2010). SRAMs are being stacked to increase memory capacity or are being used in combination with other memory parts such as flash memories and DRAMs (such as multichip package [MCP] or chip scale package [CSP]).

OEMs are embedding SRAM into different circuits (microprocessors, ASICs, flash/SRAM), thus reducing dependency on discrete SRAM parts. The PC industry, which was the largest user of SRAMs, is moving away from using off-chip SRAM cache memory in PC boards. L2 cache SRAM is embedded into MPUs, resulting in fewer discrete SRAMs being sold. ASIC logic designers are currently incorporating SRAMs into the majority of all new designs; the integration of PC L2 cache onto the microprocessor will free up SRAM wafer capacity. SRAM vendors have to find new and advanced technology solutions to free up wafer capacity.

Reducing the minimum feature sizes in semiconductor chips is enhancing the integration capabilities of SRAM circuits within other circuits, resulting in the fast growth of embedded technology.

SRAM access times are going down and supply voltages for SRAMs are going down to 1.65V and below. As stated earlier, Samsung offers SRAMs with access times as low as 0.45 ns. It has been observed that, in general, most of the SRAMs use a supply voltage of 1.8V or less, while relatively few SRAMs use 5V. Package styles used include dual in-line packages (DIPs), quad flat packages (QFPs), small outline J-leaded packages (SOJs), small outline packages (SOPs), thin small outline packages (TSOPs), ball grid arrays (BGAs), and fine-pitch ball grid arrays (FBGAs) (Hitachi, 2000; Toshiba, 2000; Samsung, 2000). Based on the trends discussed, a road map is presented in Figure 5-16. The packages used most in 2010 for SRAM have been identified as QFP, BGA, TSOP, small TSOP, and DIP.

Taking into consideration the previous forecast on sync-burst 1Mb SRAM with 3.3V in a QFP package (see Table 5-7), the zone of obsolescence was 2000, forecasted to be between 2006 and 2009. In order to validate this forecasting information, the availability for this part was researched on manufacturers' websites and is presented in Table 5-8.

Indeed, the sync SDRAM is no longer available from most manufacturers. The main reason for this unavailability is that 5 out of 15 analyzed suppliers no longer offer the required package. Moreover, 2 out of 15 suppliers no longer support 1Mb and are focused on higher densities. Another four manufacturers quit the SRAM market and focused on PSRAM, and there was one supplier

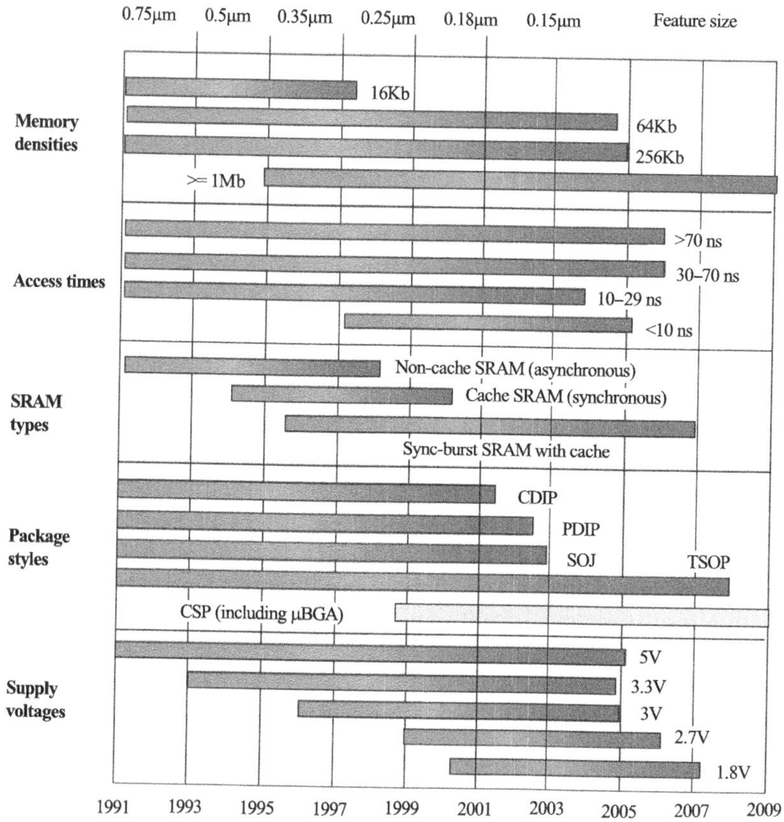

FIGURE 5-16 The history of SRAMs.

TABLE 5-8 Research Results for Sync-Burst SRAM 1Mb 1.8 V QFP in 2010

Manufacturer	Result (Reason)
Alliance	Not available (package)
ISSI	Not available (package)
AMIC	Not available (package)
Chiplus	Not available (package)
Cypress	NRND:CY7C1297H-133AXC
Etron	Not available (only PSRAM)
Fidelix	Not available (only PSRAM)
Hynix	Not available (only PSRAM)
Lyontek	Not available (package)
Micron	Not available (only PSRAM)
Renesas	Not available (density)
Samsung	Not available (EOL)
BSI	Not available (only Async)
Microchip	Not available (density)
IDT	14 devices found

who only focused on asynchronous SRAM. Finally, there were three manufacturers found who still listed the sync SDRAM on their product websites. Only Cypress still offers the requested memory, but it advises that the memory is "Not Recommended for New Designs." To sum up, there is only one manufacturer claiming that this sync SDRAM is still purchasable in a lot of variations.

As seen in the forecasting matrix in Table 5-7, the package was forecasted to become the critical element. This is in accordance with reality, because 33.33 percent of the former potential suppliers no longer offered this package in 2010.

One might argue that the forecasting result did not reflect reality in 2010, because there was still one manufacturer left. This is true, but this forecasting methodology is performed by companies who aim at using state-of-the-art devices and avoiding single-sources; and for their applications, the remaining single-source supply translates into a very high risk, which makes the memory as uninteresting for their application as if it were already obsolete.

Compared with the DRAM forecasts, the results for SRAMs are a better match to reality. The reason for this can be found in Figure 5-10 and additional historical sales data. As stated in Section 5.1.4, forecasting results are close to reality if the circumstances in the past, present, and future do not diverge. Since the amount of units shipped and sold did not fluctuate very much, the SRAM forecasting results matched reality much better.

5.3 NON-VOLATILE MEMORIES

This section presents an approach to forecasting the life cycle stage and the years to obsolescence of non-volatile memories (NVMs). This category includes EEPROMs and flash memories. Moreover, EPROMs and ROMs are included in this category as well, but because they contribute less than 1 percent of the 2006 memory shipments (McClean et al., 2007), they will be not discussed in this section. In 2006, the flash memories composed 92 percent of the non-volatile memory market (McClean et al., 2007). Because of that, this section will focus on flash memories. The approach discussed in this section is based on past sales data. A normal distribution is used along with part package style and bias level trend data to generate a forecast. Trends in the EEPROM and flash memory industry are discussed.

Memory is defined as non-volatile if it has the ability to retain its data irrespective of voltage fluctuations or the presence of a supply voltage (Christiansen, 1997). It is an advantage for the end user of a memory module to have memory that will function effectively without losing data in the event of a power failure or disconnection. Magnetic and optical media offer this advantage. Semiconductor memories have also been developed with this ability and are called non-volatile memories (NVMs). An ideal NVM is one that has low cost per bit, high density, fast random access, equal duration read-write and cycle times, low power consumption, operation over a wider range of temperatures, low voltage and power operation, and high radiation tolerance (Sharma, 1997).

5.3.1 Types of Non-Volatile Memories

The first category of NVMs is read-only memory (ROM). In ROM, data is permanently written in the manufacturing phase and cannot be altered by the user. The first integrated circuit memory was ROM, developed by Fairchild Semiconductors in 1967. ROMs used to be custom-made. The end user gave the manufacturer a truth table of the required circuit logic functions which the manufacturer converted into a ROM mask pattern that implemented the required logic functions (Sharma, 1997).

Another category of NVM, programmable read-only memory (PROM), is a ROM that can be modified once by a user (Christiansen, 1997). These are called one-time programmables (Sharma, 1997). The PROM has a matrix of memory storage cells, each of which contains a transistor diode with a fusible link in series with an electrode. PROM allows users to designate the cells that should have a 1 or 0 and enables making changes to a microcode program using a special machine called a PROM programmer. This machine supplies an electrical current to specific cells in the ROM that effectively blows a fuse in them and secures the bit pattern. The process is known as burning the PROM. Since this process leaves no margin for error, most ROM chips designed to be modified by users use erasable programmable read-only memory (EPROM) or electrically erasable programmable read-only memory (EEPROM) (Christiansen, 1997; McDaniel, 1993). The major NVMs used in the life cycle study are EEPROM and flash memory.

5.3.1.1 EEPROM In 2006, 18 percent of all worldwide memory unit shipments were EEPROM memories. EEPROM is a user-modifiable read-only memory that can be erased and reprogrammed. However, an EEPROM chip has to be erased and reprogrammed completely; selective programming of a few cells is not possible (Christiansen, 1997).

The major advantages of EEPROMs include low voltage operation, low power dissipation, high memory density, data protection lockout, wide temperature operation (military range of −55°C to 125°C), moderate access times, and moderate endurance during write cycles.

However, EEPROMs have drawbacks. They cannot perform read cycles while storing information. Longer access time requirements and endurance limitations restrict its application in certain systems. The endurance (number of write cycles) is limited because of trapped charges in the oxide, which changes transistor characteristics. Moreover, complicated manufacturing processes and complicated internal structure result in lower yield and higher cost per part (Sharma, 1997).

EEPROMs are available in either serial or parallel versions. Serial ones are mainly suited to low-cost consumer and portable applications, while parallel ones are suited to military and aerospace applications. EEPROMs are used mainly in personal programmable, low-density applications such as modems, telephone answering machines, TV controllers, set-top boxes, cell phones, and various industrial applications (McClean et al., 2007). Table 5-9 shows leading manufacturers for 2010.

TABLE 5-9 2010 Leading EEPROM Manufacturers

EEPROM Manufacturer
Atmel
STMicroelectronics
Microchip
On Semiconductor
Maxim
Holtek
NXP
Renesas
Rohm
Seiko

5.3.1.2 Flash Memory The most important category of NVMs is flash memory, which was originated by Toshiba in 1980 (Toshiba, 2010a). Flash memory is a non-volatile semiconductor memory consisting of one-transistor cells, wherein charge storage in the gate dielectric is the memory mechanism. Flash memory enables serial insertion of data and simultaneous erasure of defined blocks of memory electronically (SEMATECH, 2002). The term flash refers to the fact that the contents of the whole memory array or that of a memory block can be erased in one step.

Flash memory offers several benefits compared to other non-volatile options. It provides greater density than EEPROMs, and its non-volatility makes it a choice in some traditional RAM (volatile) memory applications. Flash memory is re-programmable, which means that OEMs can easily upgrade system software capabilities using flash. Flash memory is reasonably cost-competitive with most memory alternatives. Another major advantage of flash over traditional choices is its smaller footprint, which saves board space.

Flash memory is not as fast as RAM or ROM, but it provides greater density than EEPROM. Thus, flash memory finds applications in portable, handheld, and wireless applications. Low-density flash (1Mb, 2Mb) has found applications in PC BIOS memory; 4Mb densities can be found in digital answering machines and personal recording products, 512Mb in industrial embedded systems, and 1Gb and higher density flash find applications in telecommunications, memory cards, digital cameras, and hard-drive replacements (solid-state discs) in portable computers. Demand for applications such as set-top boxes, digital versatile disk (DVD) players, gaming, and digital cameras continues to increase the amount of flash used in consumer end applications.

Flash memory is divided in two major groups by their technology: NAND and NOR. NAND memories are commonly used in consumer electronics, whereas NOR memories are widely designed into mobile phones and industrial products. The difference between NAND and NOR technology leads to different endurances. For example, NAND provides about 20 times more

TABLE 5-10 2010 Leading Flash Manufacturers

Flash Manufacturer
Samsung
Toshiba
Micron
Spansion
ISSI
EON
AMIC
Winbond

TABLE 5-11 Non-Volatile Memories: A Comparison (Karls et al., 1998; Brown and Brewer, 1998)

Characteristic	Flash	EEPROM
Typical storage use	In-system modifiable programs	Frequently updated programs and data
Typical number of writes	Write up to 100,000 times	Write up to 1 million times
Densities available	4Mb to 64Gb	1Kb to 512Mb (serial) 16Kb to 4Mb (parallel)
Flexibility	Medium	Most
Cost per bit	Medium	Most

read-and-write cycles than equivalent NOR. Nevertheless NOR memory has faster access times (random-access-reading) but lower densities compared to NAND. However, NAND, due to its smaller structures, offers larger densities with less speed (page-access only).

The leading flash memory suppliers for 2010 are shown in Table 5-10. A comparison of EEPROM and flash memory is presented in Table 5-11.

5.3.2 The Non-Volatile Memory Market

Figure 5-17 shows the sales trend of flash memory, and the EEPROM market is shown in Figure 5-18.

High-volume memory markets, such as DRAM, SRAM, and flash, are driven by low cost per bit and high memory density (Karls et al., 1998). However, because of byte-by-byte program/erase abilities and very low power consumption, serial EEPROMs find their own niche application in cost-sensitive and power-limited mobile communications and portable electronic products. In applications where 1Kb through 512Mb is enough memory, EEPROMs offer a flexible and affordable solution (Karls et al., 1998). Even though the use of EEPROMs might save money, they have already begun fading out (the comparison of typical ASPs in Figure 5-17 and Figure 5-18 shows that common EEPROMs only cost about one-tenth as much as flash memory).

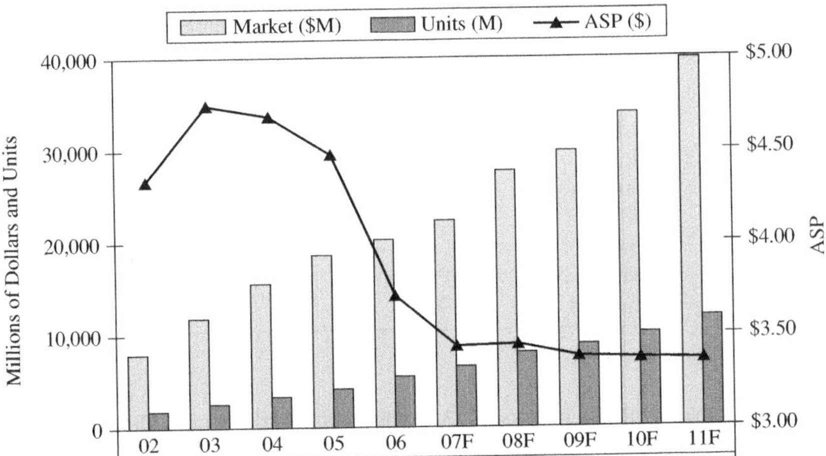

	02	03	04	05	06	07F	08F	09F	10F	11F
Market ($M)	7,767	11,739	15,611	18,569	20,275	22,364	27,588	29,791	33,980	39,739
Units (M)	1,791	2,473	3,331	4,145	5,455	6,499	7,993	8,804	10,069	11,824
ASP ($)	4.34	4.75	4.69	4.48	3.72	3.44	3.45	3.38	3.37	3.36

FIGURE 5-17 Flash memory sales, shipments and ASPs.
Source: WSTS, IC Insights

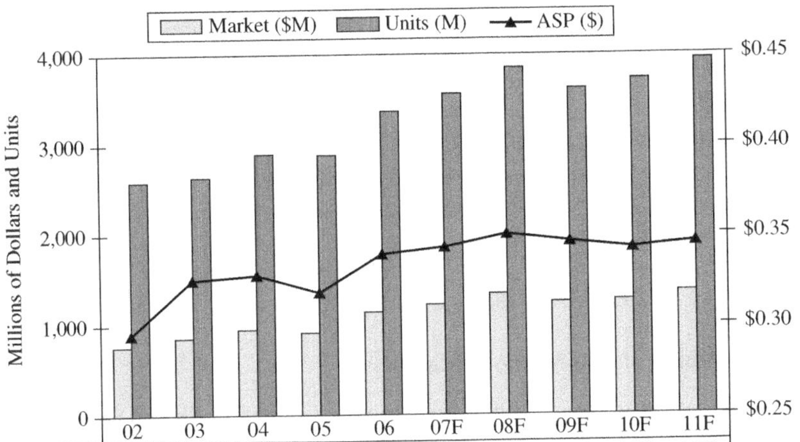

	02	03	04	05	06	07F	08F	09F	10F	11F
Market ($M)	763	857	950	915	1,140	1,220	1,342	1,248	1,273	1,362
Units (M)	2,587	2,636	2,893	2,875	3,360	3,560	3,839	3,606	3,715	3,936
ASP ($)	0.29	0.33	0.33	0.32	0.34	0.34	0.35	0.35	0.34	0.35

FIGURE 5-18 EEPROM sales, shipments, and ASPs.
Source: WSTS, IC Insights

5.3.3 Application of the Life Cycle Forecasting Methodology

To forecast both the years to obsolescence and the life cycle stage, we use a model based on sales figures. A primary attribute determines the curve-fit of the sales data to the model, and modifying attributes translate the obsolescence prediction depending on their degree of influence over the primary attribute. In the case of EEPROMs and flash, the primary attribute is the memory size, and the dominant modifying attributes are package style and nominal supply voltage. Because of these attributes, the evolutionary approach is applicable. It is up to the desired confidence level to decide if the Basic Life Cycle Method (see Section 4.1.1) or the Advanced Life Cycle Method (see Section 4.1.2) should be applied. In this case, the Advanced Life Cycle Method will be applied, because the flash market faces a higher-than-average volatility, which requires forecasts with a high level of confidence.

Figure 5-17 and Figure 5-18 show sales trends of EEPROMs and flash memory.

To perform a forecast the following equation (refer to Section 4.1.2) is applied:

$$Obsolescence\ Window = \mu_P + (\mu_{1o} \pm x\sigma_{1o})\sigma_P \tag{5-1}$$

In the next step, it is required to get μ_P and μ_{1o} as well as σ_P and σ_{1o}. The mean and the standard deviation for the sales peak will be calculated the same way, as done in the Basic Life Cycle Method previously.

To determine these parameters, historical shipment or sales data are required. This data can be found in Figure 5-19.

Using this shipment data, a trend line was developed by using the method of least squares and the following equation:

$$\mu(x) = c\ln(x) + b \tag{5-2}$$

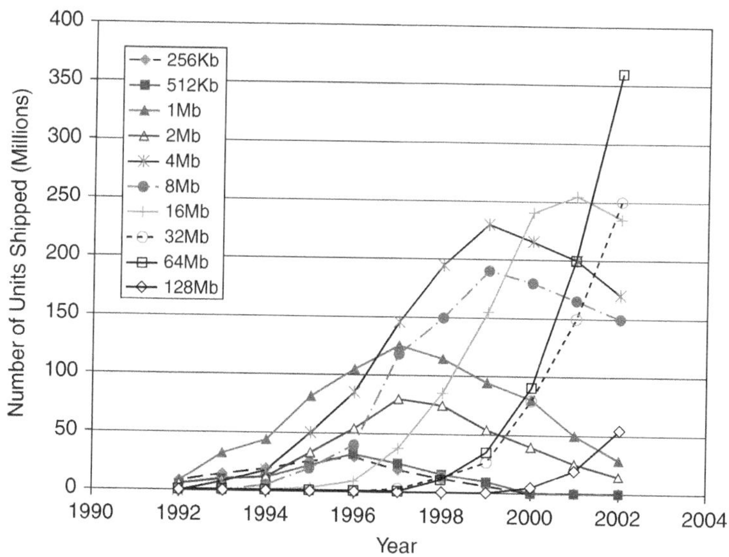

FIGURE 5-19 Historical shipment data for flash memory by density (Sandborn et al., 2007a).

The parameter x represents the density in Mb and μ the peak shipment year. The method of least squares delivers c and b. As a result, the trend equation used to fit the curve is

$$\mu_P = 1.5663\ln(x) + 1997.2 \qquad (5\text{-}3)$$

In the next step, the standard deviation of the Gaussian is fit to the sales data.

$$\sigma_P = -0.0281\ln(x) + 2.2479 \qquad (5\text{-}4)$$

Equations 5-3 and 5-4 represent a Gaussian model created according to the Basic Life Cycle Method. To achieve the Advanced Life Cycle Method, the level of confidence x and the manufacturer-specific obsolescence probability factors μ_{1o} and σ_{1o} must be determined.

As stated, these factors are manufacturer-specific and therefore cannot be assumed in general. Historical data with a manufacturer is required to determine the window of obsolescence for that manufacturer's parts.

As described in Section 4.1.2, the time period between the peak sales year of the part (calculated with the Basic Life Cycle Method) and the actual discontinuance date issued by the manufacturer has to be normalized using standard deviations (calculated with the Basic Life Cycle Method). This has to be done for every part instance for the selected part type and manufacturer.

After this examination, the normalized discontinuance dates can be plotted in a histogram, which represents a probability distribution. Again, using Gaussian fit, the mean μ_{1o}, as well as the standard deviation σ_{1o}, can be computed. The example in Figure 5-21 was discussed in Section 4.1.2.

According to Figure 5-20, the mean μ_{1o}, as well as the standard deviation σ_{1o}, are determined as follows:

$$\mu_{1o} = 0.88$$
$$\sigma_{1o} = 0.72$$

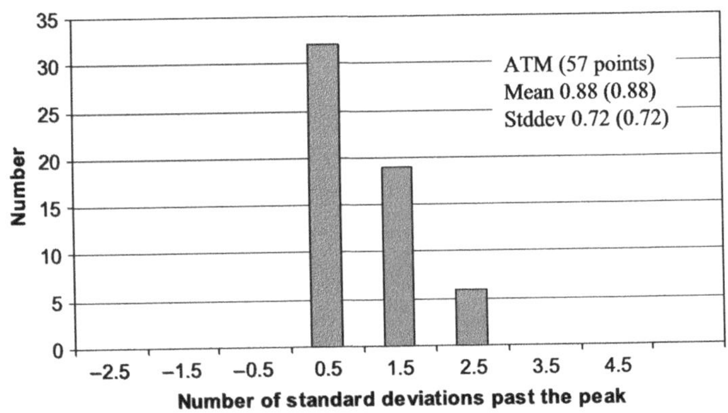

FIGURE 5-20 Atmel flash memory last order dates projected in standard deviations past the peak.

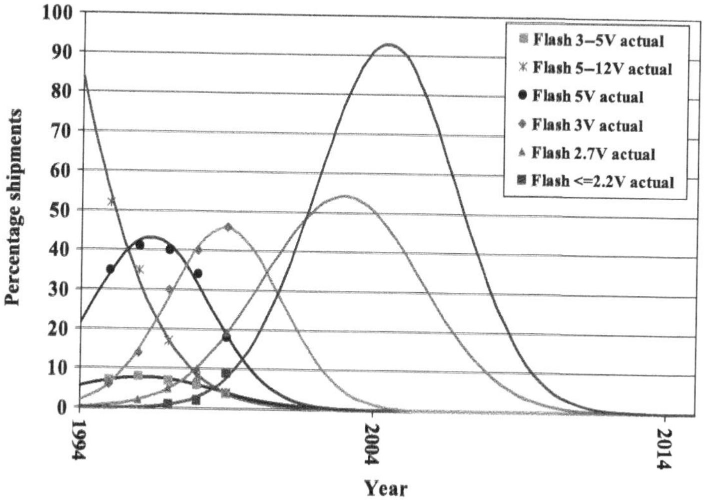

FIGURE 5-21 Flash memory life cycle curves by read, erase, and program voltage.

With a desired confidence level of 91.1 percent, the factor x is chosen to be 1.5. To summarize, Equation 5-5 with all the factors is given by

$$\begin{aligned} \textit{Obsolescence Window} = {} & 1.5663\ln(x) + 1997.2 + (0.88 \pm 0.72 \cdot 1.5) \\ & \times (-0.0281\ln(x) + 2.2479) \end{aligned} \tag{5-5}$$

To get a specialized forecast for an Atmel flash, the memory size x has to be entered. To get a forecasting result, Atmel's active DataFlash memory AT45DB161D will be used. This flash memory provides 16Mb and 8 I/O pins with a supply voltage between 2.7 V and 3.6 V. This memory was last updated to revision N in May 2010 (Atmel, 2010c). The device is available in TSOP 28, SOIC 28, BGA, and MLF (VDFN) packages. It is the successor of the AT45DB161 and AT45DB161B, which are already discontinued.

Following Equation 5-5, the window of obsolescence is forecasted as described in Table 5-13. It is essential not to forget the secondary parameters, in this case defined as supply voltage and package.

Life cycle curves for flash memory by voltage are shown in Figure 5-21. The effects of modifying attributes may shift the years to obsolescence and the life cycle stage of the main attribute. See Table 5-12.

5.3.4 Determining the Zone of Obsolescence

If the years to obsolescence for any of the modifying attributes fall within the life span of the main attribute, the years to obsolescence for the generic part will be modified. The algorithm is summarized in Chapter 4 (Table 4-3), where the

TABLE 5-12 Life Cycle Parameters for EEPROM and Flash Memory by Read, Erase, and Program Voltages, Analysis Date $p = 1999$

Memory Type	Voltage (V)	Mean (μ)	Std. Dev. (σ)	$\mu+2.5\sigma-p$	$\mu+3.5\sigma-p$
EEPROM	5	2010.5	10.5	37.8	48.4
Flash read, erase, and program voltage	5–12	1989.6	3.3	– 1.0	2.3
	5	1996.4	2.0	2.4	4.4
	3-5	1996.1	2.5	3.4	5.9
	3	1998.9	1.9	4.7	6.6
	2.7	2002.9	2.7	10.7	13.4
	≤ 2.2	2004.4	2.4	11.4	13.8

TABLE 5-13 Advanced Life Cycle Forecasting Method Performed on Atmel's AT45DB161D Flash Memory with 16Mb Density in 2010

Attribute	μ (μ_j for modifying attributes)	$\sigma(\sigma_j$ for modifying attributes)	$(\mu + 2.5\sigma - p, \mu + 3.5\sigma - p)$	Comparision of limits	Modified zone of obsolescence	Comments
Main: Megabit size (16Mb)	2001.5	2.2	(−3, −0.8)			
Modifying: Package style (TSOP)	1999.7	5.2	(2.7, 7.9)		(min [−3, 2.7], min [−0.8, 7.9]) = (−3, −0.8)	No effect
Modifying: Voltage (3.3V)	2002.9	2.7	(−0.35, 2.35)		(min [−3, −0.35], min [−0.8, 2.35]) = (−3, −0.8)	No effect

modifying attributes are denoted by j, with μ_j and σ_j denoting the mean and standard deviation of the modifying attribute. Based on one of the conditions given in Chapter 4 (Table 4-3), a modified zone of obsolescence, given by the ordered pair (a, b), can be determined. The ordered pair (a_{min}, b_{min}) that is the smallest in numeric value between the remaining ordered pairs (a, b) is the modified zone of obsolescence.

5.3.5 Discussion of Non-Volatile Memory Forecasts

In general, ROM and EPROM markets have been replaced partly by EEPROM, but mainly by flash memory. There may still be some niche applications using ROM or EPROM parts. This observation was made from past market history; the EPROM market collapsed from a $1.4 billion market in 1995 to $515 million in 1997 (Karls et al., 1998). The reason for this sharp decline was that the average selling price (ASP) for flash memory dropped significantly and designers turned to the more versatile flash memory for their systems.

Data storage applications such as cellular phones, network equipment, telecom switches, and printers accounted for 57 percent of the $20 billion flash memory market (Karls et al., 1998) in 2006. The NAND flash segment especially is driven by the consumer and communication sector.

Since automotive and industrial applications only accounted for 3 percent of the 2006 total flash market, it is obvious that manufacturers' focus is influenced by three main segments (consumer 16 percent, communications 41 percent, and computer 40 percent) (McClean et al., 2007). Because of the immense dynamics in these three segments, the other segments, due to their long product life cycles, might face early obsolescence problems.

The trend in most IC manufacturers is to embed memory on board a logic chip, thereby producing a system-on-a-chip solution; in many cases, flash memory is the memory of choice for mating. At present, manufacturers offer embedded flash alongside microcontrollers, digital signal processors, or even alongside SRAM or analog circuitry. This provides OEMs with greater design freedom, lower system costs, and more board area for today's space-constrained applications. Flash memory, along with microcontrollers and digital signal processors, reduces chip-to-chip latency, resulting in faster access times for flash. Flash vendors are working on increasing the density of embedded flash and lowering programmable voltages. In embedded microcontroller applications, flash is currently replacing EEPROM. See Figure 5-22.

Many companies are integrating flash memory chips to gain memory capacity and better functionality. For example, the ColdFire+ 32-bit microcontrollers by Freescale Semiconductors in 2010 offered an integrated 90 nm thin film storage (TFS) flash technology with FlexMemory. This microcontroller provides high-performance, mixed-signal capabilities, ultra-low-power capabilities, and high precision (Freescale, 2010c).

The FlexMemory technology provides the following features (Freescale, 2010d):

- Flex memory is a memory block that can be configured as EEPROM and flash memory.
- Combination of TFS flash array (FlexNVM) + SRAM block (FlexRAM) + hardware state machine.
- Fully embedded in the MCU—avoids software complexity and eliminates system resource impact on CPU/flash/RAM of EEPROM emulation schemes.

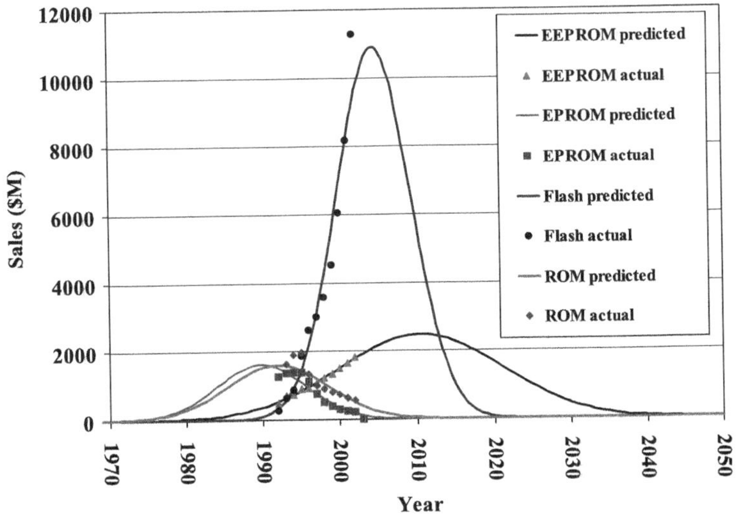

FIGURE 5-22 Life cycle curves of EPROM, EEPROM, flash, and ROM.

- Can be accessed simultaneously with the main program memory.
- Can be used for storing additional application program code, data tables, or byte write/erase system data.

By 2010, the FlexMemory technology only provided 32Kb up to 128Kb, but due to its 90 nm structures, there is room for further evolution. For example, Samsung Semiconductors announced on October 13, 2010, that the production of flash memory in 20 nm structures had started (Samsung, 2010d).

Leaving these general trends aside, at first glance the forecasting results in Table 5-13 might surprise. According to the forecasting results, Atmel's DataFlash AT45DB161D should no longer be available in December 2010 or become obsolete until October 2011. The memory was introduced in October 1997 and focused on applications such as PC BIOS memory.

Taking a closer look at the revision history, the initial version, AT45DB161, was available from October 1997 to April 2001. Its successor, the AT45DB161B, originally introduced in 2000, became obsolete in January 2006. The current version, AT45DB161D, was still available in 2010. The latest datasheet revision of this device is dated May 2010 (Atmel, 2010b; 2010c).

One might argue that the discontinuance now might be forecasted too early, but if so, there is one very important aspect left aside. The statistical framework of the Advanced Life Cycle Method is based on declining sales and shipments. Once sales and shipments reach a critical level, the manufacturer consolidates the product. This might result in discontinuance, product improvement, or die shrinking.

In this case, the consolidation led to the introduction of BGA packages in April 2009. Moreover, the sector and chip erase time was reduced by about 40 percent in May 2010. Finally, the cumulative page erase/program was doubled as well. According to Table 5-13 this consolidation was forecasted between 2007 and 2010.

To sum up, the Advanced Forecasting Method did forecast a window in which product discontinuance was highly likely. But instead of making this product obsolete, the manufacturer decided to expand the product life cycle. Moreover, it has to be pointed out that AT45DB161D focuses on the PC Bios market, which does not have the same high volatility as other flash market segments. By using the forecasting methodology, one must be aware of the fact that this information is only statistical.

In general, the future of non-volatile memories is definitely flash memory. If manufactured with the right technology, such as stacked technology or embedded technology, flash memory is the undisputed leader of non-volatile memory.

5.4 MICROPROCESSORS

This section presents the application of the life cycle algorithm to forecast the life cycle stage and the years to obsolescence of microprocessors. A microprocessor is a computer processor on a microchip. A microprocessor is designed to perform arithmetic and logic operations that make use of small number-holding areas called registers. There is a clear difference between microprocessors and microcomputers. Microcomputers offer on-chip memory and can execute programs directly. In comparison, the microprocessor requires additional components (such as memory).

A microprocessor typically consists of the following (Harper, 1999):

1. Arithmetic logic unit (ALU) to perform integer operations such as addition, subtraction, multiplication, and division.
2. Floating point unit (FPU) to perform the same operations on floating-point numbers.
3. Numeric processors to perform trigonometric, exponential, and double-precision calculations.
4. Graphic processors to display input, output, and images on the screen.
5. Memory in the form of cache and registers.
6. Interface logic and communication on the bus.

In 1971, Texas Instruments invented the single-chip "calculator on chip," later called the first microcomputer (Texas Instruments, 2010a). In the same year, Intel Corporation introduced the first single-chip microprocessor, the 4004 device, under a contract with a calculator company. Since then, microprocessors have become the basis of the world's most powerful computers. At the other end of the spectrum, microprocessors can cost as little as 50 cents and use only microwatts of power (the 45 nm Intel Atom Z500 only requires 650 mW [Intel, 2010c]).

Intel is currently the largest microprocessor manufacturer in the world. There is a long list of other companies (see Table 5-14) that have found success for their microprocessors in various markets due to the inexorable expansion of opportunities and applications for microprocessors.

Although everyone agrees that microprocessor performance will continue its dramatic rate of increase in the future, there are many different opinions about how it will achieve that goal. Microprocessor performance can be increased both by improving the manufacturing process and by changing the architecture of the chip. First, improvements in manufacturing process bring a natural increase in microprocessor speed. The world's most advanced microprocessors in 2010 were manufactured in a 32 nm process by Intel Corporation (Intel, 2010d). In the near future, however, microprocessor manufacturing geometries should fall below that level using new technologies, such as hafnium-based high-k metal gate (HKMG) technology. For instance, in August 2010 Toshiba announced that a new 24 nm process had been implemented on certain memory products (Farrell, 2010). Of course, memory has different technological requirements than microprocessors, but it shows that technological evolution as stated by Gordon Moore has not come to an end so far.

Microprocessors are available in 8-bit, 16-bit, 32-bit, and 64-bit designs. The largest market subcategory combines 32-bit and 64-bit microprocessors. In 2010, the 32-bit systems were still dominant, but due to there being no changes to the Windows 7 sales price, the 64-bit designs are gaining market share. This is mainly because it allows users to apply 4GB or even more memory (Farrell, 2010). Intel has dominated the 32-bit market since the advent of x86 microprocessors.

TABLE 5-14 Manufacturers of Microprocessors in 2010

Intel
AMD
VIA Semiconductors
IBM
Freescale
Texas Instruments
Microsemi
Fujitsu
Rabbit
Renesas
PMC Sierra
Integrated Device Technology
MIPS Technologies
Analog Devices
ST Microelectronics
Marvell

5.4.1 Types of Microprocessors

The microprocessor market segment consists of two main architectures: complex instruction set computing (CISC) and reduced instruction set computing (RISC). A comparison of RISC and CISC architectures is shown in Table 5-15.

The most commonly used microprocessors are CISC; however, RISC microprocessors dominate the performance-minded microprocessor market. Despite RISC's historical performance advantage, it has failed to put a dent in the Intel PC monopoly. RISC still controls some important computing niches. Nevertheless, the IBM Microelectronics/Motorola PowerPC RISC microprocessor finally lost the high-volume computing market in 2005 when Apple announced that it would focus on Intel-based systems in the future (Rentrop, 2010).

5.4.2 The Microprocessor Market

For the overall microprocessor market, the x86 microprocessor is the dominant sector, and Intel enjoys a dominant market share of more than 80 percent. Figure 5-23 shows the leading vendors' microprocessor market shares in 2010.

Intel is still the largest of the CISC microprocessor suppliers. When plotted logarithmically, the number of transistors has increased along an upward slope for the past 35 years, as shown in Figure 5-25. Still, RISC architectures are used in certain high-volume applications. For instance, the Sony Playstation 3 uses IBM's Cell Broadband Engine, which is closely linked to the RISC architecture.

Figure 5-24 shows that the majority of processors are utilized to meet the demands of computer and communication applications.

To qualify the market's strong progress, the number of transistors per MPU and the year of introduction are shown in Figure 5-25.

Using this transistor count data, a fit for the number of transistors n in year x was developed that uses the following form:

$$n(x) = ae^{bx} \tag{5-6}$$

which becomes

$$n(x) = 89278e^{(x-1970)0.9263} \tag{5-7}$$

TABLE 5-15 Characteristics of RISC versus CISC Microprocessors

Feature	CISC	RISC
Instruction set size	Optimized	Reduced and growing
Compiler technology	Needs optimization	More efficient than CISC
Memory requirements	–	20 to 30% more than CISC
Performance	–	1/2 to 1 speed generation ahead of CISC
Price	Equivalent to RISC	Equivalent to CISC
Market share	Dominant	–
Application	Mainly consumer	Mainly industry

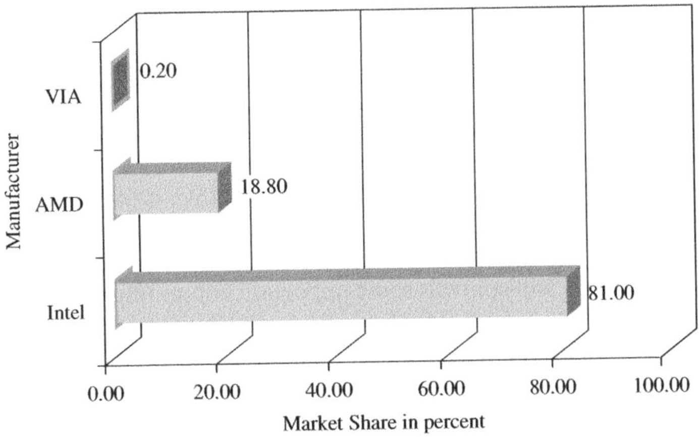

FIGURE 5-23 Overall MPU market share in percent of total MPU market in 2010 (LePedus, 2010).

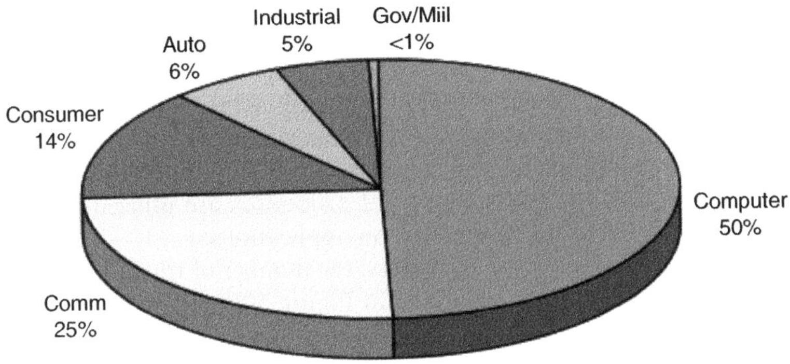

FIGURE 5-24 2006 MPU market by system application (McClean et al., 2007). *Source:* WSTS, IC Insights

Furthermore, the coefficient of determination was determined:

$$R^2 = 0.97967$$

The evolution of Intel and AMD processors is described in Table 5-16.

5.4.3 Application of Forecasting Methodology

To forecast both the years to obsolescence and the life cycle stage, we use a model based on sales or shipment figures. A primary attribute determines the curve-fit to the sales or shipment data to the model, and modifying attributes translate the obsolescence prediction depending on their degree of influence over the primary attribute. Microprocessors have strong evolutional parametric drivers. Therefore, the Life Cycle Forecasting Methodology is applied. In

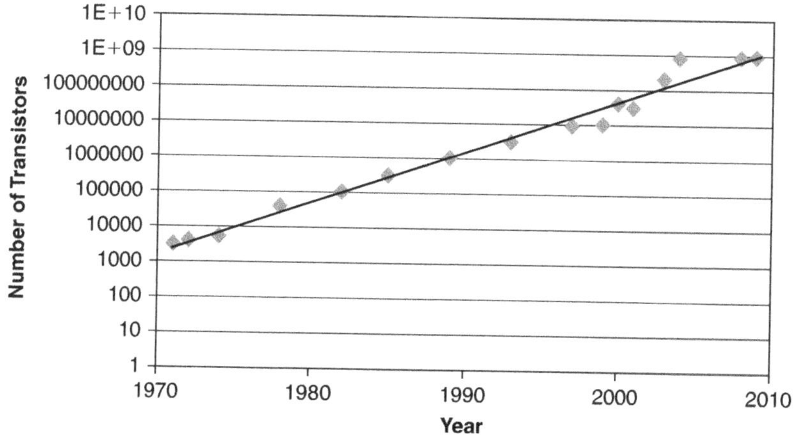

FIGURE 5-25 Number of transistors versus time (Intel, 2010e).

order to provide a clear overview of this approach with regard to microprocessors, the Basic Life Cycle Forecasting Methodology is chosen in this section. It is up to the analyst to decide on their requirements for the forecasting process to apply the advanced approach as well.

In the case of microprocessors, the primary attribute used to be microprocessor speed. In recent years, there has been a slight drift in this evolution. Microprocessor speed and density used to be the key indicators for performance until multi-core microprocessors were finally introduced. By combining the performance power of several microprocessor cores, the total performance can be improved significantly. Since increasing clock frequencies lead to more power consumption in general, multi-core access enables more performance with less energy. As far as densities are concerned, the structures continue to shrink. Therefore, multi-core microprocessors eclipsed the classic single-core processor.

Today's primary attributes are clock speed, number of cores, bus width, and density. Power consumption can be counted as a primary attribute as well, but since the acceptable level of power consumption varies strongly from system to system, this section refers to it as a secondary attribute.

Because there are clear parametric evolutionary drivers for microprocessors, the Basic Life Cycle Method (see Section 4.1.1) or the Advanced Life Cycle Method (see Section 4.1.2) should be applied.

Figure 5-26 shows a plot between microprocessor bus width and the corresponding year of peak sales. Since the majority of shipped microprocessors in computing in 2010 were multi-core 64-bit processors, the sales peak was interpreted as 2010. Currently, there is no visible trend that 64-bit systems will become obsolete in the next few years. Nonetheless, it is highly recommended to update the following equations with the latest sales peak each year. Using the sales data, a trend equation was derived.

TABLE 5-16 Microprocessor Development History (Mann, 2000; Intel, 1998, 1999, 2010e; Toronto Star, 1996)

Year	Microprocessor Products
1971	Intel 4004 microprocessor, 2300 transistors
1972	Intel 8008 microprocessor, 3300 transistors
1974	Motorola 6800, 4000 transistors
	Intel 8080 microprocessor, 8-bit, 2 MHz, 6000 transistors
1975	MITS Altair 8800 with 256 bytes of random access memory
1978	Intel 8086 16-bit, 29000 transistors, 5, 8, and 10 MHz
1979	Intel 8088, 8-bit, 29,000 transistors
	Motorola 68000
1982	Intel 80286, 16-bit processor with 134,000 transistors
1984	Motorola 68020, 2,000,000 transistors, 32-bit processor
1985	Intel 80386 DX, a 32-bit processor with 275,000 transistors
1987	Motorola 68030
1988	Intel 80386 SX
1989	Intel 80486 DX, 32-bit processor with 1.2 million transistors
	AMD's copies of 386 appeared
1990	Motorola 68040 SX
1993	Intel Pentium, 3.1 million transistors, 64-bit operation
	Cyrix, AMD, IBM Microelectronics and Texas Instruments 486
1994	Motorola and IBM Microelectronics PowerPC RISC processor
1995	Intel Pentium Pro, 64-bit processor, optimized 32-bit operation, 5.5 million transistors
	AMD's K5; Cyrix's 586
1996	Intel Pentium II
1997	AMD K6, Cyrix M2
1998	Intel Pentium III, Xeon
	AMD K6 III, PowerPC G3
1999	AMD Athlon; Intel Pentium III Kaimai
2000	AMD Thunderbird, Duron
	Intel Coppermine
2001	AMD Mustang
	Intel Itanium
2003	AMD K8 series
	Intel Itanium 2
2004	AMD K8 series
	Intel Itanium (9MB cache)
2006	AMD K8 series
	Intel Core 2
2007	AMD K10 series
	Intel Dual Core
2008	AMD K10 series
	Intel i7
2009	AMD K10 series
	Intel Xeon

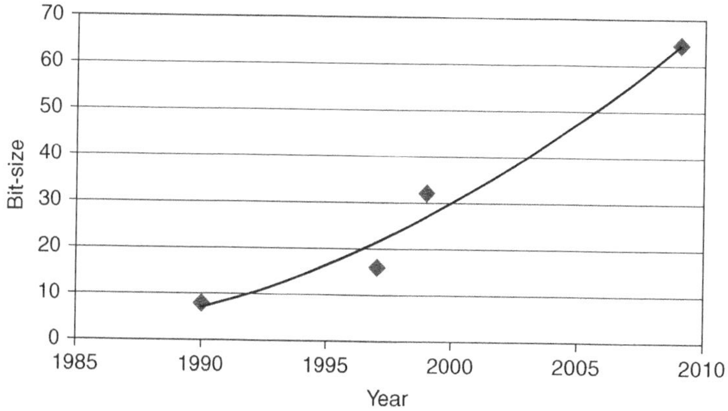

FIGURE 5-26 Peak sales year versus microprocessor size.

Using the bit count data from Figure 5-26, a trend line for the number of bits in the year x was developed by using:

$$T(x) = ax^2 + bx + c \tag{5-8}$$

which becomes

$$T(x) = 0.0819x^2 - 324,55x + 321530 \tag{5-9}$$

Furthermore, the coefficient of determination was determined:

$$R^2 = 0.96975$$

Figure 5-28 shows a plot of the microprocessor bus width and the standard deviation of the Gaussian fit to the sales data. The different phases in the life cycle of the part are defined by the standard deviation σ values.

Maturity is defined as two σ centered about the mean (that is, the year in which the sales reach a maximum). Decline, phase-out, growth, and introduction each occupy one σ portion of the remaining life cycle curve on either side. The zone of obsolescence is defined as the distance between the (2.5σ, 3.5σ) points measured from the maximum sales of the part μ. Knowing the date of assessment, one can determine a range for the years to obsolescence of the generic part, given as (μ + 2.5σ date of assessment, μ + 3.5σ date of assessment). Its life cycle stage can be determined from the curve list zones of obsolescence for microprocessors based on bit size and architecture. Figure 5-27 and Figure 5-28 show the life cycle curves. See also Table 5-17.

In the case of microprocessors, there are secondary attributes such as supply voltage and packaging that may modify the life cycle stage or the years to obsolescence.

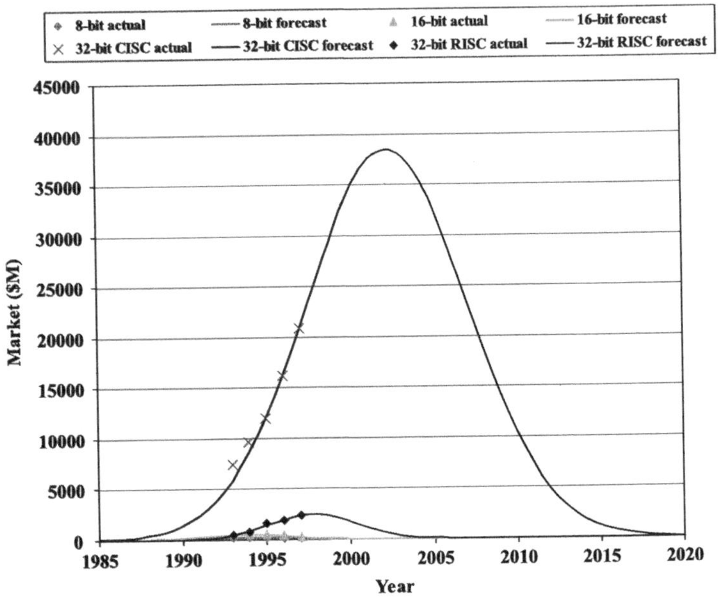

FIGURE 5-27 Life cycle curves of microprocessors based on bit-size.

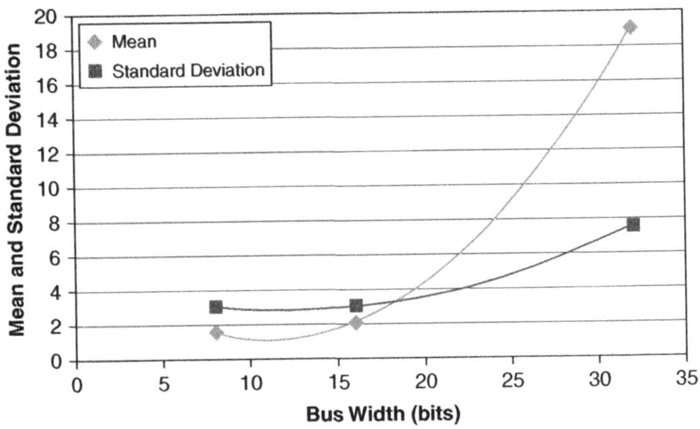

FIGURE 5-28 Mean and standard deviation versus microprocessor bus width.

TABLE 5-17 Zones of Obsolescence Based on Bit-Size (p = Date of Assessment)

Microprocessor Bit-Size	μ	σ	$\mu+2.5\sigma-p$	$\mu+3.5\sigma-p$
8-bit	1989.7	5.6	4.2	9.8
16-bit	1996.5	4.3	7.6	11.9
32-bit	1999.8	3.6	9.4	13.0
64-bit	—	—	—	—

5.4.4 Determining the Zone of Obsolescence

The case studies chosen deal with the following Intel microprocessors: Pentium (without MMX), Pentium Pro, Pentium (with MMX), and Pentium II (with MMX). The mean μ and standard deviation σ for each of these part categories are listed in Table 5-18. Figure 5-29 shows the life cycle curves of the Intel part categories listed in Table 5-18. With the increase of microprocessor functionality or complexity, the percentage shipment for the specified type is decreased; additionally, the life cycle becomes shorter.

5.4.5 Discussion of Microprocessor Forecasts

The trend of netbook personal computers has had an impact on microprocessor markets. In applications where low cost, low power consumption, and low package height are the requirements, CPUs are still built in the CISC x86 microprocessor architecture. But for ultra-mobile devices, such as tablets and mobile phones, special CPUs are used. For example, Apple's iPad uses an

TABLE 5-18 Mean μ and Standard Deviation σ Values for Intel's Microprocessors

Intel Part Type	σ	μ
Pentium (without MMX)	0.5	1996.6
Pentium Pro	0.5	1997.1
Pentium (with MMX)	0.49	1997.8
Pentium II (with MMX)	0.48	1998.5

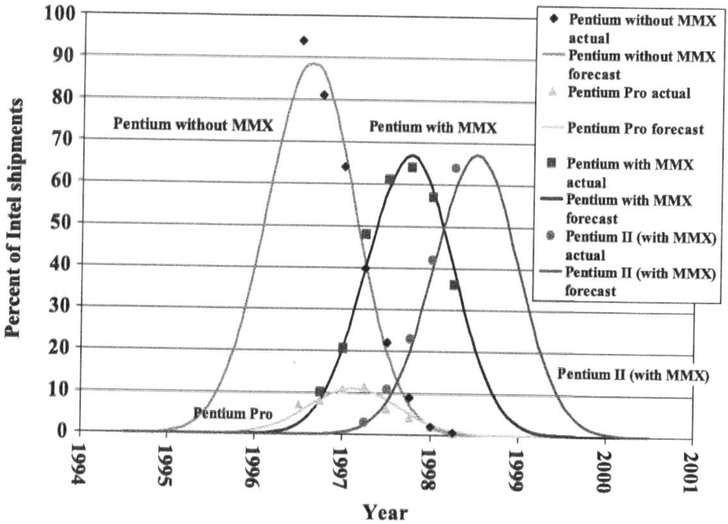

FIGURE 5-29 Intel microprocessor life cycle curves.

application-specific CPU designed by PA and manufactured by Samsung with part number K4X2G643GEJGC8 (Ross, 2010).

As stated earlier, the trend is to combine multiple CPUs into one in order to improve performance. Another trend for ultra-mobile devices is to use application-specific CPUs.

Intel has invested billions of dollars on wafer fabrication capabilities and state-of-the-art processing equipment to reach and maintain its position as a world semiconductor leader. Intel has started to manufacture most of its processors using 32 nm or smaller process technology. By reducing the process geometries used to manufacture its microprocessors, Intel is able to keep manufacturing costs under control and resist the efforts of its competitors to gain market share.

Currently, 64-bit advanced six-core microprocessors, such as Intel's i7, are manufactured in 32 nm technology, and AMD's Phenom II x6 is manufactured in 45 nm technology. Intel's i7 is working at 3.3 GHz clock frequency, and AMD's Phenom II x6 at 3.2 GHz.

Figure 5-30 shows the historical trends for package styles, transistor count, data bus width, and architecture. The supply voltage for Intel's i7 is at about 1.4V, and AMD's Phenom requires about 1.3V. Clock frequencies reach about 3.4 GHz, which is triple the frequency common in 2001.

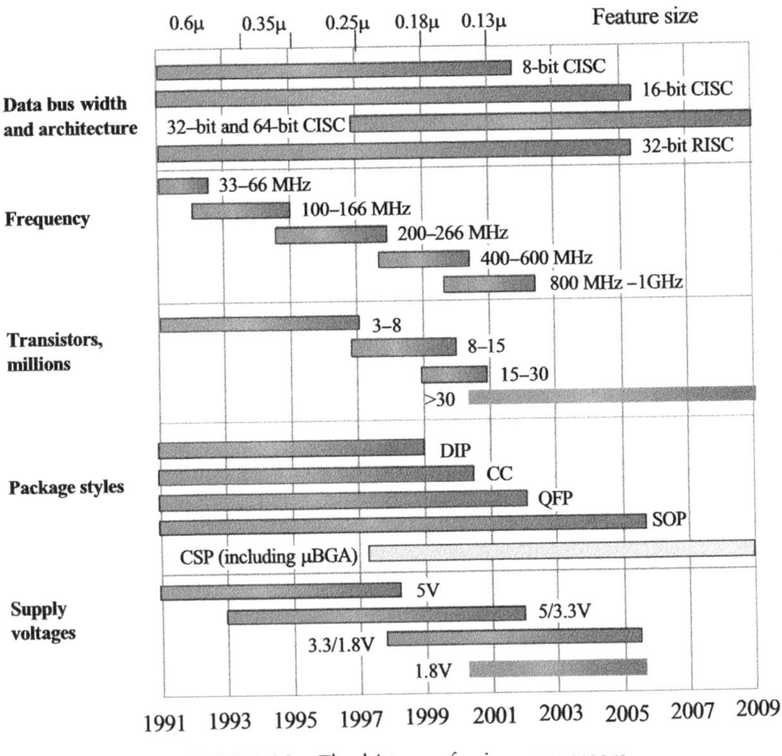

FIGURE 5-30 The history of microprocessors.

5.5 MICROCONTROLLERS AND DIGITAL SIGNAL PROCESSORS (DSPs)

This section presents an approach for forecasting the life cycle stage and the years to obsolescence of microcontrollers and DSPs. A microcontroller is a highly integrated chip that contains all the components comprising a controller.

As defined by the WSTS, a microcontroller is a stand-alone device that performs a dedicated or embedded computer function within an electronic system without the need of other support circuits (McClean et al., 2007). Typically, a microcomputer includes a CPU, RAM, some form of ROM, I/O ports, and timers. Unlike a general-purpose computer, which also includes all of these components, a microcontroller is designed for a very specific task—to control a particular system. Microcontrollers are sometimes called embedded microcontrollers. They are a part of an embedded system—that is, one part of a larger part or system. Microcontrollers are primarily used in control-oriented applications that are sensing and controlling external events. In 1971, Texas Instruments invented the single-chip "calculator on chip," later called the first microcomputer (Texas Instruments, 2010a). Digital signal processors (DSPs) are a subcategory of microcontrollers, which are traditionally found in systems that require the precision processing of analog signals (Giovino, 2001).

Microcontrollers can be used in any part that measures, stores, controls, calculates, or displays information to control objects, processes, or events. Currently, there are several microcontrollers used for engine controls in cars. In desktop computers, microcontrollers are inside soundcards, keyboards, modems, printers, and other peripherals. In test equipment, microcontrollers make it easy to add features such as the ability to store measurements, create and store user routines, and display messages and waveforms. Consumer products that use microcontrollers include cameras, video recorders, MP3 players, and ovens. At the same time, new applications are emerging, such as wireless communication equipment, Voice over Internet Protocol (VoIP), and Internet download audio and broadband asymmetric digital subscriber lines (ADSL). These applications result from the high-speed development of digital technology and the Internet. Table 5-19 provides a list of common microcontroller applications.

Microcontrollers represent about 10 percent of all semiconductor sales (McClean et al., 2007). As both the Internet and advanced process technologies bring new applications, microcontrollers are expected to become an even more significant part of the IC market. It is forecast that the microcontroller market is expected to grow more than 10 percent annually (McClean et al., 2007).

5.5.1 Type of Microcontrollers

Microcontrollers can be broken down into three groups:

1. Embedded microcontrollers.
2. External memory microcontrollers.
3. Digital signal processors (DSPs).

TABLE 5-19 Typical Microcontroller Applications

Automobile	Consumer	Internet and Communication
Board computers	Answering machines	Third-generation (3G) wideband cellular phones
Engine controls	Computers	Pagers
Air bags	Home entertainment—A/V systems	Voice over Internet Protocol (VoIP)
Instrumentation	Camcorders	Gateway systems
Security systems	Electronic musical instruments	Audio players
Transmission controls	Digital cameras	Modems
Entertainment	Exercise equipment	Remote access concentrators
Climate control	Smoke detectors	Broadband asymmetric digital subscriber lines (ADSL)
Keyless entry	Microwaves	Smart antennas for base stations
Navigation	Telephones	
Collision avoidance systems	Fax machines	
	Copiers	
	Laser printers	
	Organizers/PDAs	
	Watches	
	Smart cards	
	Washing Machines	
	Coffee Machines	

5.5.1.1 Embedded Microcontrollers In an embedded microcontroller, all the necessary resources, such as memory and I/O, are built into the chip, except the parts that provide power and clocking. These microcontrollers can be based on an established microprocessor core or use a design specific to the embedded application. The primary role of these microcontrollers is to provide programmable logic control and interfacing with the external world.

5.5.1.2 External Memory Microcontrollers Most notable parts with 16- or 32-bit data paths rely completely on external memory. This external memory contains the control unit, as well as any RAM, which allows customers to change application software. It is the most obvious difference between the embedded microcontroller and the external memory microcontroller.

5.5.2 The Microcontroller Market

The microcontroller market can be defined by 4-bit, 8-bit, 16-bit, 32-bit, and DSP configurations. In 2006, the microcontroller/DSP market grew 4 percent (McClean et al., 2007). The fastest-growing category is 32-bit microcontrollers, and the largest category is 8-bit. This will not be true much longer, considering the intense pressure from other categories. Figure 5-31 shows the 2006 microcontroller/DSP market distribution.

The tradeoff between performance and price is a major concern for the selection of microcontrollers. The automotive and consumer electronics sectors drive the growth of the microcontroller IC market and are characterized by the low cost of 4-bit and 8-bit microcontrollers. MCU usage by the electronics sector is shown in Figure 5-32.

The dramatically declining cost of 32-bit parts and the pressure put on OEMs to differentiate their products based on performance have driven the rise in high-end microcontrollers in applications such as video games. It is forecasted that there will be a big jump from 8-bit to 32-bit, with only a short period of transition to 16-bit because of its address limitations (Mayer, 2000).

An alternative to the microcontroller is the digital signal processor (DSP). The capability of DSPs to have real-time programmable signal processing satisfies the requirements of the digital age with complete systems-on-a-chip. Three factors are largely responsible for driving the demand for DSPs. First, the need for bandwidth expansion; second, the need for higher quality and precision control; and third, the need to future-proof or provide seamless upgrade ability.

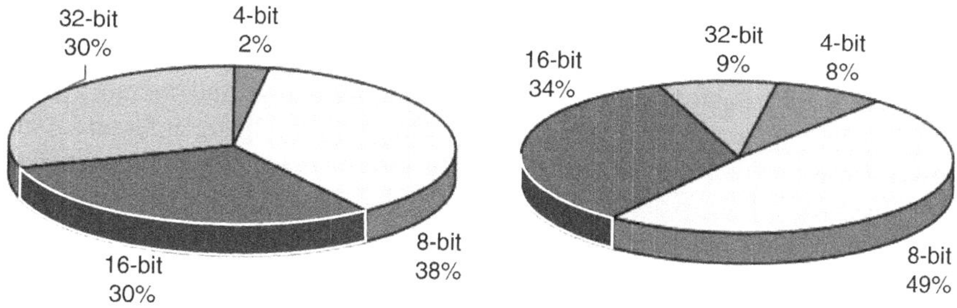

FIGURE 5-31 2006 microcontroller/DSP market distribution (McClean et al., 2007). *Source:* WSTS, IC Insights

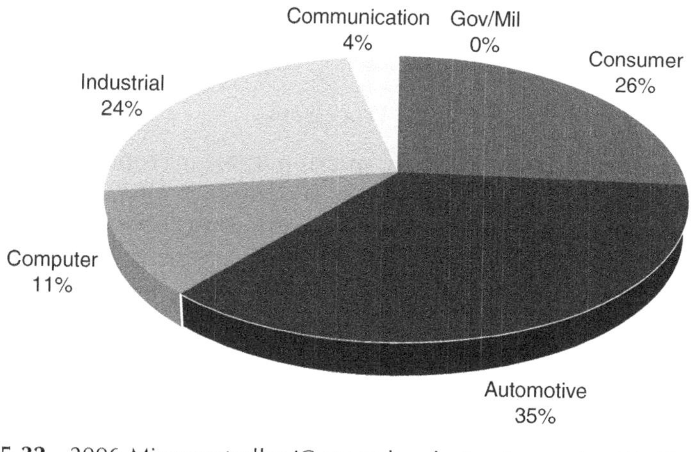

FIGURE 5-32 2006 Microcontroller IC usage by electronics market (McClean et al., 2007).

TABLE 5-20 2008 Microcontroller Market (Riemenschneider, 2010a)

Company	Market Share
Renesas Electronics	29.8%
Freescale	11.0%
Fujitsu	7.7%
Infineon	7.2%
Microchip Technology	5.9%
STMicroelectronics	4.7%
Texas Instruments	4.4%
Atmel	3.7%
NXP Semiconductors	2.1%
Others	23.5%

Table 5-20 shows the 2008 leaders in microcontroller sales. Texas Instruments sold its volatile DRAM business in May 1998 to Micron Technologies and made DSPs its core competency. In December 2010, Texas Instruments announced that their focus would shift to analog semiconductors and stand-alone application processors for smart phones (IT-Times, 2010).

As customer requirements become more sophisticated and memory prices continue to fall, microcontroller cores are increasingly being integrated with various types and densities of memory. Embedded processors incorporating flash and other types of non-volatile memory is a major trend, which makes it easy for customers to change or modify microcontrollers to their individual requirements. Microcontrollers with embedded DRAM are also being well received, since they can provide an excellent solution for applications such as digital cameras, printers, and tablets that require image processing and high-speed access between the microprocessor and memory. A higher level of integration is another trend; for example, integrating microcontroller cores and DSPs with other peripheral blocks for smartphones. These integrated semiconductor devices are referred to as stand-alone application-specific processors for smartphones.

5.5.3 Overview of Digital Signal Processors

Digital signal processors (DSPs) are special-purpose microprocessors designed to handle signal-processing applications very quickly. DSPs and their ALUs (arithmetic logic units—the hardware that performs calculations) take sample data from an analog system and calculate an appropriate response at very high speed to allow real-time control (Predko, 1999).

5.5.4 Application of the Life Cycle Forecasting Methodology

To forecast both the years to obsolescence and the life cycle stage, we use a model based on sales figures. Since microcontrollers show evolutional tendencies regarding bit-size, the Basic Life Cycle Method as well as the Advanced

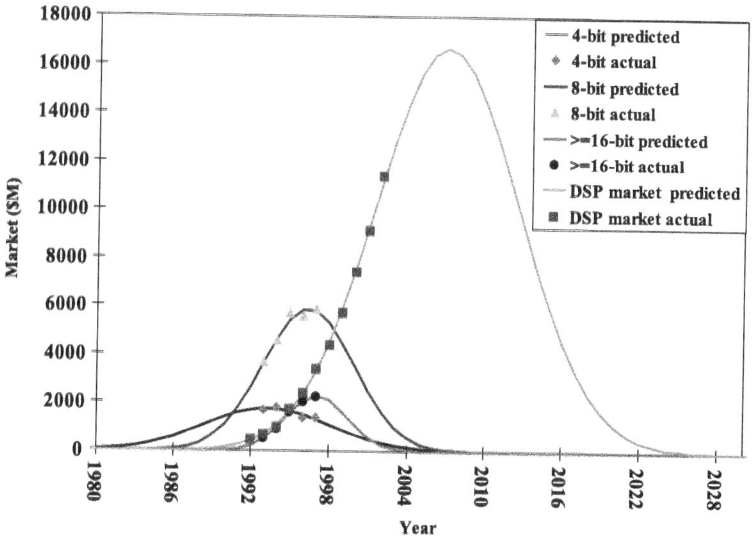

FIGURE 5-33 Life cycle curves for microcontrollers and DSPs.

TABLE 5-21 Mean μ and Standard Deviation σ Values for Microcontrollers and DSPs

Part Type	μ	σ
4-bit	1993.3	4.9
8-bit	1996.4	3.5
$\geq$ 16-bit	1997.1	2.4
DSP	2003	5.5

Life Cycle Method can be applied. The primary attribute determines the curve-fit to the sales data, and the modifying attributes modify the obsolescence prediction depending on their degree of influence over the primary attribute. In the case of microcontrollers, the primary attribute is the bit-size. We have used data from the ICE (Karls et al., 1998) to develop an algorithm to forecast the life cycle stage and years to obsolescence of a microcontroller part. The results of the forecast can be seen in Figure 5-33 and Table 5-21.

5.5.5 Determining the Zone of Obsolescence

If the years to obsolescence for any of the secondary attributes fall within the life span (three σ points) of the main attribute, the years to obsolescence for the generic part will be modified. The algorithm is summarized in Chapter 4, where the secondary attributes are denoted by j, with μ_j and σ_j denoting the mean and standard deviation of the secondary attribute. Based on the applicable condition given, a modified zone of obsolescence, given by the ordered pair (a, b), can be determined. The ordered pair (a_{min}, b_{min}) that is the smallest in numeric value between the remaining ordered pairs (a, b) obtained from the conditions given is the modified zone of obsolescence.

5.5.6 Discussion of Microcontroller and DSP Forecasts

Using the values of mean μ and standard deviation σ obtained from curve fits from both microcontrollers and DSPs, their life cycle curves may be drawn. The mean μ and standard deviation σ for each of these part categories are listed in Table 5-21. The curve fits for the microcontroller and the DSP categories are shown in Figure 5-33.

Figure 5-33 shows that 4-bit microcontrollers followed a decline out of market after reaching a peak in 1993. The 8-bit part sales peaked in 1996; the 16-bit part reached its peak in 1997.

High-end microcontrollers include 16-bit and 32-bit parts. The market share of 16-bit parts is being penetrated by 8-bit parts due to their lower price; at the same time, 32-bit parts are beginning to replace 16-bit because of their address limitations. With 122 case study forecasts and future trends from the Internet and home appliances, Mitsubishi Electronics expects a jump from 8-bit to 32-bit microcontrollers. This trend can also be seen in Figure 5-31.

In summary, there is a clear trend to combine as many components and as much functionality into one microcontroller/DSP as possible.

5.6 LOGIC PARTS

This section presents an approach to forecast the life cycle stage and the years to obsolescence of logic parts. Logic parts and their families cannot be described simply as memories, microprocessors, or microcontrollers because they do not have a unique, clear evolutional driver. Moreover, the families, also referred to as types of logic parts, are closely connected to the manufacturer's individual product portfolio strategy. Further, a detailed analysis of every single logic part in complex systems is not cost-effective at all. Therefore, the application of procurement lifetime (see Chapter 4) is reasonable. The obsolescence dates for logic parts could be forecasted using the Basic or Advanced Life Cycle Method if an evolutionary parameter can be identified.

5.6.1 Types of Logic Parts

Logic parts are digital integrated circuits that control the operation of electronic systems and move data (Fairchild, 2000). Logic parts such as bus drivers, transceivers, clock drivers, gates, shift registers, multiplexers, and decoders perform a multitude of functions and are classed into families depending on technology type, such as CMOS, BiCMOS, ECL, and TTL.

A sampling of the major features of different logic families (Texas Instruments, 2000g) is provided next.

ABT (Advanced BiCMOS Technology): The ABT family is manufactured using a 0.8 μm BiCMOS process and provides high drive up to 64 mA and propagation delays below the 5 ns range while maintaining very low power consumption. To reduce transmission line effects, the ABT family has series-damping resistor options. Further, there are special ABT parts that provide extremely high current drive (180mA) for use with 25 Ω transmission lines.

AC/ACT (Advanced CMOS Technology): The AC/ACT family of parts is manufactured in 1 μm CMOS and has more than 160 part types, including gates, latches, flip-flops, buffers/drivers, counters, multiplexers, transceivers, and registered transceivers. The AC/ACT family is a low-power logic family with 24 mA output current drive at 5V V_{CC} and 12 mA output current drive 3.3V V_{CC}. Output-edge control (OEC) circuitry, available only with the center-pin products, helps reduce simultaneous switching noise associated with high-speed logic. AC parts offer CMOS-compatible inputs, and ACT parts offer TTL-compatible inputs.

AHC/HCT (Advanced High-speed CMOS/High-speed CMOS): The AHC/AHCT logic family provides a natural migration for HCMOS users who need more speed for low-power, low-noise, and low-drive applications. Performance characteristics of a typical AHC family (Texas Instruments) are:

1. Speed: With typical propagation delays of 5.2 ns (octals), which is about three times faster than HC parts, AHC parts are the quick and quiet solution for higher speeds at 5V V_{CC} operation.
2. Low noise: The AHC family allows designers to combine the low-noise characteristics of HCMOS parts with today's performance levels without the overshoot/undershoot problems typical of higher-drive parts usually required to get AHC speeds.
3. Low power: The AHC family, by using CMOS technology, exhibits low power consumption (half that of HCMOS).
4. Drive: Output-drive current is ± 8 mA at 5V V_{CC} (nominal supply voltage) and ± 4 mA at 3.3V V_{CC}.

ALB (Advanced Low-Voltage BiCMOS Technology): The 3.3V ALB family uses 0.6 m BiCMOS process technology for bus-interface functions. The ALB family provides 25mA drive at 3.3V with maximum propagation delays of 2.2ns, making it one of the fastest logic families to date.

ALS (Advanced Low-Power Schottky): The ALS family, combined with the AS family, can be used to optimize systems through performance budgeting. By using AS in speed-critical paths and ALS where speed is less critical, designers can optimize speed and power performance. The ALS family includes gates, flip-flops, counters, drivers, transceivers, registered transceivers, readback latches, clock drivers, register files, and multiplexers.

ALVC (Advanced Low-Voltage CMOS Technology): One of the highest performance 3.3V bus-interface families is the ALVC family. The 3V products are processed in 0.6 m CMOS technology, giving typical propagation delays of less than 3 ns, along with current drive of 24 mA and static power consumption of 40 μA for bus-interface functions. ALVC is becoming the industry standard for all 3.3V logic applications. The family also features innovative functions that make it ideal for memory interleaving, multiplexing, and interfacing to SDRAMs.

ALVT (Advanced Low-Voltage BiCMOS Technology): ALVT is a high-performance 3.3V bus-interface logic family. The 5V-tolerant 3.3V products use 0.6 μm BiCMOS technology for bus-interface functions and delivers 2.4 ns propagation delays, a current drive of 64 μA, and static power consumption of 90 μA. Some ALVT families also provide series-damping resistors to reduce transmission line effects.

AS (Advanced Schottky): The AS family of high-performance bipolar logic offers high-drive capabilities. This family, combined with the ALS family, can be used to optimize system speed and power through performance budgeting. By using AS in speed-critical paths and ALS where speed is less critical, designers can optimize speed and power performance. The AS family includes gates, flip-flops, counters, drivers, transceivers, registered transceivers, read-back latches, clock drivers, register files, and multiplexers.

BCT (BiCMOS Technology): BCT is a family of drivers, latches, transceivers, and registered transceivers. Designed specifically for bus-interface applications, BCT offers TTL I/O with high speeds and low power in the disabled mode. The parts in the BiCMOS low-impedance driver family can maximize the speed and reliability of heavily loaded systems.

CBT (Crossbar Technology): In today's computing market, power and speed are two of the main concerns. CBT can address both of these issues in bus-interface applications. CBT enables a bus-interface part to function as a very fast bus switch, effectively isolating buses when the switch is open and offering very little propagation delay when the switch is closed. These parts can function as high-speed bus interfaces between computer-system components, such as the central processing unit (CPU) and memory. CBT parts also can be used as 5V to 3.3V translators, allowing designers to mix 5V or 3.3V components in the same system.

CBTLV (Low-Voltage Crossbar Technology): CBTLV parts can be used in multiprocessor systems as fast bus connections, bus-exchange switches for crossbar systems, ping-pong memory connections, or bus-byte swapping. They also can be used to replace relays, improve connect/disconnect speed, and eliminate relay-reliability problems. The CBTLV family has been generally designed to operate at 3.3V.

GTL (Gunning Transceiver Logic Technology): GTL technology is a new reduced-voltage switching standard that provides high-speed, point-to-point communications with low power dissipation. Companies such as Texas Instruments offer GTL/TTL translators to interface with TTL-based subsystems. This enables designers to use the GTL switching standards for speed-sensitive subsystems and to use the translators to interface with the rest of the system. GTL parts often have innovative circuitry, such as bus hold, on the inputs to eliminate the need for external pull-up resistors for floating inputs, which reduces power, cost, and board-layout time. Output edge-rate control (OEC) is offered on the outputs to reduce electromagnetic interference (EMI) caused by the high frequencies of GTL.

HC/HCT (High-Speed CMOS Technology): HC/HCT logic is especially suited for low-power logic requirements. Over 100 part types, including gates, latches, flip-flops, buffers/drivers, counters, multiplexers, transceivers, and registered transceivers fall in this category. The HC family offers CMOS-compatible inputs, and the HCT family offers TTL-compatible inputs.

LV (Low-Voltage CMOS Technology): LV CMOS technology products are specially designed parts for 3V power supply use. Some manufacturers such as Texas Instruments have recharacterized the LV family to operate at 5V.

LVC (Low-Voltage CMOS Technology): LVC logic products are specially designed for 3V power supplies. Texas Instruments' LVC family is a high-performance version with 0.8 μm CMOS process technology, 24 μA current drive, and 6.5 ns maximum propagation delays for driver operations.

LVT (Low-Voltage BiCMOS Technology): The 3.3V LVT family uses 0.8 μm BiCMOS-process technology with performance specifications ideal for workstations, networking, and telecommunication applications. Performance characteristics of a typical LVT family (Texas Instruments) are:

1. Speed: Provides high performance with typical propagation delays of 4 ns for buffers.
2. Voltage: Provides 3.3V operation with 5V tolerant I/O and capability to interface with a mixed-voltage environment. The I/Os can handle up to 7V, which allows them to act as 5V/3V translators.
3. Drive: The LVT family provides up to 64 μA of drive, yet consumes less than 100 μA of standby power at 3.3V V_{CC}.
4. Live insertion: LVT parts also incorporate circuitry to protect the parts in live-insertion applications. The parts go into the high-impedance state during power up and power down, which is called powered-up 3-state (PU3S).

74F (Fast Logic): 74F logic is a general-purpose family of high-speed advanced bipolar logic. Parts, including gates, buffers/drivers, bus transceivers,

TABLE 5-22 **Characteristics of Common Logic Families (Texas Instruments, 2000g)**

Process Technology	High Speed	Low Noise	Low Static Power	High Drive	Low Voltage	Board Space
BiCMOS 5V	ABT			ABT		
BiCMOS 3V	LVT			LVT	LVT	
CMOS	GTLP	GTLP			GTLP @ 3.3V	
CMOS 5V	TinyLogic UHS	FACT QS	FACT			TinyLogic HS
		HC/HCT	FACT QS			TinyLogic HST
		VHC/VHCT	HC/HCT			TinyLogic UHS
		VHC/VHCT	HC/HCT			TinyLogic UHS
			VHC/VHCT			
CMOS 3V		LVX	LCX		LCX	TinyLogic HS
			LVX		LVX	TinyLogic UHS
	TinyLogic UHS		VCX		VCX	
Bipolar	FASTr	ALS		FASTr		
		FAST				
ECL	300 Series ECL					

TABLE 5-23 Procurement Lifetimes for Various Electronic Part Types through 2008

	Censored				Uncensored					
Part Type	Mean (years)	β	η	*LKV*	Mean (years)	β	η	*LKV*	% of Parts Not Obsolete	Total Number of Parts
Linear regulators	11.63	2.842	13.06	−1205	8.245	3.470	9.168	−822	59.46%	509
Buffer & line drivers	38.39	2.021	43.33	−3042	9.821	4.008	10.83	−1293	91.08%	1279
Bus transceivers	15.39	2.289	17.37	−1746	9.281	4.988	10.11	−984	57.26%	1057
Decoder & demux	20.74	1.713	23.25	−914.9	9.298	4.505	10.19	−490	62.30%	565
Flip flop	16.23	2.250	18.33	−1727	9.638	5.154	10.48	−950	58.27%	1052
Inverter Schmitt trigger	18.13	1.830	20.40	−1125	8.575	4.061	9.453	−627	64.13%	750
Latch	14.99	2.289	16.92	−1391	9.186	4.952	10.01	−792	55.46%	818
Multiplexer	18.83	1.819	21.19	−844.0	8.835	4.096	9.734	−470	63.70%	552

(β and η refer to 2-parameter Weibull fits of the censored and uncensored PDFs. *LKV* is the negative log-likelihood function (larger negative values indicate a better fit). (Sandborn et al., 2011)

flip-flops, latches, counters, and multiplexers/demultiplexers, are in the 74F logic family.

5.6.2 The Logic Part Market

The logic part market in 2006 was worth \$150M. Most of the advanced logic families are offered by major manufacturers in packaging, such as the shrink small-outline package (SSOP) and thin shrink small-outline package (TSSOP) to ensure higher performance and reduced board space. Table 5-22 shows the characteristics of some common logic families.

On the PC motherboard, logic ICs work as "glue chips" for performing diverse functions such as bus drivers, clock drivers, switches, and gates. In cellular phones, logic ICs are used as transceivers. As silicon products become faster, smaller, and more cost effective, most electronic parts, such as processors, memories, and ASICs, run at lower voltage. Low-voltage logic functions need converted voltages.

5.6.3 Application of Forecasting Methodology

The procurement lifetime approach is chosen. Since this approach has been already performed in Chapter 4, there will be no detailed description in this chapter. For a detailed description of this approach, please refer to Section 4.2. The results of this approach can be found in Table 5-23.

In order to compare the purchasing lifetime results with the Basic Life Cycle Methodology, a short assessment of the Basic Life Cycle approach was performed, and the results are shown next.

Table 5-24 shows the values for the mean μ and standard deviation σ for the logic part technologies, together with zone of obsolescence.

Table 5-25 shows the life cycle stages of the various logic families manufactured by different semiconductor companies. The primary factors of importance here are the mean μ and standard deviation σ, as these factors determine the years to obsolescence and the life cycle stage of the logic IC. The life

TABLE 5-24 Mean μ and Standard Deviation σ Values for Logic Technologies

Logic Family	μ	σ	$\mu+2.5\sigma-p$	$\mu+3.5\sigma-p$
NMOS	1982.2	5.1	−4.2	0.9
PMOS	1966.5	6.6	−16.0	−9.4
CMOS	2005.3	12.5	37.5	50.0
BiCMOS	1997.2	3.2	6.1	9.3
S/LS TTL	1978.4	8.3	0.0	8.3
TTL	1964.5	10.4	−8.6	1.8
ECL	1979.3	7.4	−1.3	6.1

(Date of assessment $p=1999$ [Karls et al., 1998])

TABLE 5-25 Life Cycle Stages for Various Logic Families of Different Manufacturers, Their Manufacturing Technologies and Operating Voltages (Karls et al., 1998; Texas Instruments, 2000g)

Life Cycle Stage	Introduction	Growth	Maturity	Decline	Phase-out	Discontinuance
Technology	Diamond, SOI	GaAs	CMOS, BiCMOS	ECL	S/LS TTL, TTL, NMOS	PMOS
Years to obsolescence	Data not available	Data not available	CMOS: 37–50 yrs BiCMOS: 6–9 yrs	ECL: 0–6 yrs	S/LS TTL: 0–8.3 yrs TTL: 0–1.8 yrs NMOS: <1 yr	PMOS: 0 yrs
Logic family	54ALB, 54CBTLV, 54ALVT VCX, GTLP, Bus Switch, ALVC, 74CBTLV, 74ALB, 74ALVT, 7SZ	54CBT, LVX, LVT, 54LVC, 54LVQ, 54LV, HS/HST/ UHS, 54VHC, VHCT, SCAN, 74LVU, 74LVXC, 74LVC, 74LV, 15HSC, 54ETL, 54FB, 64C, 64CT, 74ETL, 74FB, 74ALVC, 74LCX, 74LPT, 74UH, 7ST, 74LCT, 74FST, GA1000	54BCT, FACT/FACT QS, 54AHC, 54AHCT, 54AHCU, 100LVEL, 29C, 49C, 54C, 4S, 54ABTE, 54ACQ, 54ACS, 54ACTQ, TQ1000, TC7W, TC7S, TC4S, TC3W, 54GTL, 54HSC, 54HST, 54QST, 64BCT, 6B, 70C, 72C, 74CT, 74C, 74BCT, 74GTL, HCTS, 74VCX, 74VHCT, 74VHCU, 80C, ACS, 74BC, 74CBTD, ABT, 54ABT	49FCT, 10KH, 100K, ECLinPS, ECL, Lite, 74F, CD4000/74C 54AS, 100LVE, 100S, 1000, 5000, 100E, 100EL, 100H, 14000, 1600, 4500, 100S, 10E, 10H, 10EL, 10K, 74AS, HCS, CMOS/ TTL, Bipolar, 74VHC, ACT, AC, ACTS, AS	HC, HCT, TTL, LSTTL, FAST/ FASTr/554ALS, 82, 54, 100K, 62, 54HCU, 54L, 54S, 64ALS, 64B, 74ALS, 74H, 74HC, 74HCT, 54HC, 54HCT, 7S, 74HCU, 74L, LS, HC, F, 74F, 74SSTL, 82C, 82S, 8T, 93L, 93S, 96L, 96LS, 96S, 74FR	71, 73, 29, 75, 74, 54, 4300, 4000, 4700, 96, 93, 85, 84, 74QST, 11C, 29FCT, 39C, 54FBT, 54H, 74FBT, 78L, 88L, 8TS, 74HL, 74B
Voltage	1.8V	2.0V, 2.5V	3.0V, 3.3V, 3.3/5.0V	5.0V, 5.2V	6.2V, 9V, 12V	> 12V
Years to obsolescence	14–17	9–12, 5–8	2–6	0–2		

cycle curves of logic ICs are modified by taking into account the effects of two modifying attributes: package style and voltage. For each of these categories of modifying attributes, the same methodology can be applied and values for the mean μ and standard deviation σ for each of the categories of modifying attributes can be obtained.

5.6.4 Discussion of Logic Part Forecasts

Figure 5-34 shows the life cycle curves of the major logic fabrication technologies. In recent years, the use of MOS logic, including ASICs, has increased in the communications and computer segments. GaAs, another emerging technology, is a niche alternative to silicon in certain high-frequency applications. CMOS is the dominant process used in making logic ICs. Older technologies, such as NMOS, PMOS, and bipolar, are being discontinued. Emitter coupled logic (ECL) technology, after maintaining about 5 to 8 percent of the logic market, is being displaced by CMOS. This is because the excellent performance characteristics of ECL technology, such as speed and output drive capabilities, are now being replicated in CMOS.

In the 1970s, bipolar (TTL logic) was the major technology of choice. BiCMOS offers IC designers a mix of the characteristics of CMOS and bipolar, and ICs can be designed with the best characteristics for each part of the circuit. However, the process complexity of BiCMOS requires more wafer fabrication steps and has kept it from rising in the semiconductor industry.

Nevertheless, it has to be mentioned that there are applications that require BiCMOS logic parts. The advantage of BiCMOS is that it can deliver a high

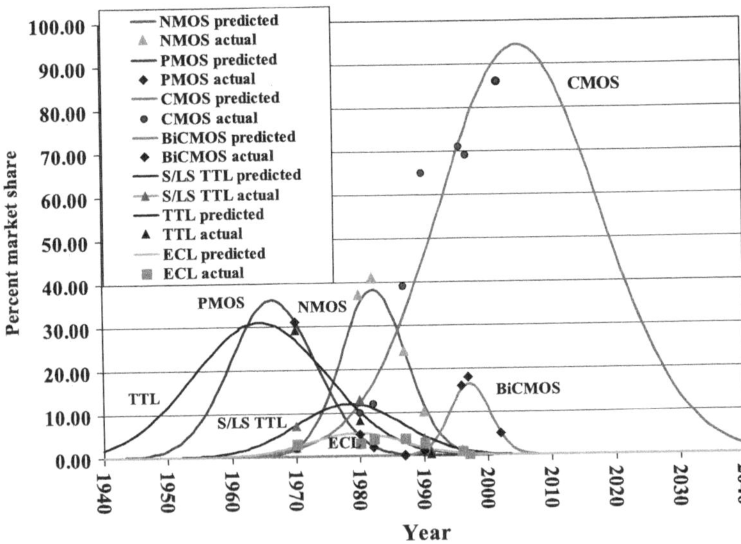

FIGURE 5-34 Life cycle curves by logic technology type.

TABLE 5-26 Basic Life Cycle Analysis on SN74BCT374DBR, $p = 2010$ (Date of Assessment)

Attribute	$\mu(\mu_j$ for modifying attributes)	$\sigma(\sigma_j$ for modifying attributes)	$(\mu + 2.5\sigma - p, \mu + 3.5\sigma - p)$	Zone of Obsolescence
Main: BiCMOS	1997.2	3.2	(−3, −0.8)	
Package style (SSOP)	1999.7	5.2	(2.7, 7.9)	(2.7, 7.9)
Voltage (5V)	1990.4	4.3	(−8.9, −4.6)	(2.7, 7.9)

current and is less sensitive to capacitive loads. In order to compare the forecasting results of the procurement lifetime forecast and the basic life cycle approach, the 8-bit flip-flop SN74BCT374DBR by Texas Instruments was chosen. The suffix "BCT" in this part number stands for this technology. See Table 5-26.

5.7 ANALOG PARTS

This section presents an approach to forecast the life cycle stage and the years to obsolescence of analog parts. There are two types of integrated circuits: digital and analog. Digital circuits process on-off electrical signals that represent the binary states 1 and 0. Analog ICs process continuously varying voltages and currents that represent continuously variable values such as temperature, pressure, speed, acceleration, position, and rotation, to name a few. Even though the trend is to digitize as much as possible, the digital world is based upon analog basics. For instance, MP3 players need digital/analog converters and amplifiers to power analog headphones. In general, there is no digital without analog.

Basically, like logic parts and their families, analog parts cannot be described as simply as memory, microprocessors, or microcontrollers, because they do not have a unique clear evolutional driver. Furthermore, a detailed analysis of every single analog part in complex systems is not cost effective. Therefore, the application of procurement lifetime (see Chapter 4) is reasonable. This does not imply, however, that logic parts cannot be forecasted with the Basic or Advanced Life Cycle Method in general.

5.7.1 Types of Analog Parts

Analog ICs can be divided into two subcategories. The first category, called standard analog, includes all amplifiers, interface ICs, voltage regulators and references, data conversion ICs, and comparators. The second category, referred to as application-specific analog, contains consumer, telecom, automotive, industrial, and computer. Telecom itself is divided into wireless and wired sectors.

Amplifier ICs amplify voltage or current output; the magnitude of amplification represents the ratio of the output voltage or current to input voltage or current. Examples of amplifier ICs are operational amplifiers, instrumentation amplifiers, buffers, video amplifiers, and radio frequency ICs. CMOS amplifiers have advantages in energy conservation and die size but have been higher priced than bipolar amplifiers. This has prevented them from gaining popularity compared with bipolar amplifiers.

Voltage regulators and reference ICs maintain voltage in a circuit at a specified level and are used to protect sensitive electronic equipment from voltage surges. Voltage regulators find applications in a wide range of electronic equipment, ranging from car alarms to cable TV control boxes. Reference ICs are "electronic benchmarks" that allow the voltage of a part to be compared against an established standard.

Interface ICs serve as intermediaries to transfer signals between or within electronic systems. Examples of interface ICs are line drivers, receivers, keyboard encoders, and display drivers.

Data-converter ICs are mixed-signal circuits containing both analog and digital circuitry. Data-converter ICs convert analog signals to digital or vice versa and are also referred to as A-to-D and D-to-A converters. Comparator ICs produce a digital output by comparing two input voltages.

The second category, application-specific analog parts, contain circuits designed specifically for usage in certain equipment such as entertainment circuits, air-bag and antilock-brake circuits, radios, televisions, and personal and home appliances. Advanced systems such as direct broadcast satellite receivers for television set-top boxes, global positioning satellite systems, and wireless local area networks are growing application areas of analog ICs. As shown in Figure 5-35, application-specific analog parts were the most frequently used analog ICs in 2006.

5.7.2 The Analog Part Market

The analog IC market has always been more stable than the digital IC market. The digital IC market has traditionally been viewed as more exciting because it represents the leading edge of semiconductor technology. The analog market represents steady and modestly profitable business to certain semiconductor manufacturers. As a result, most analog ICs are less prone to obsolescence compared with the digital IC market. Growing digital

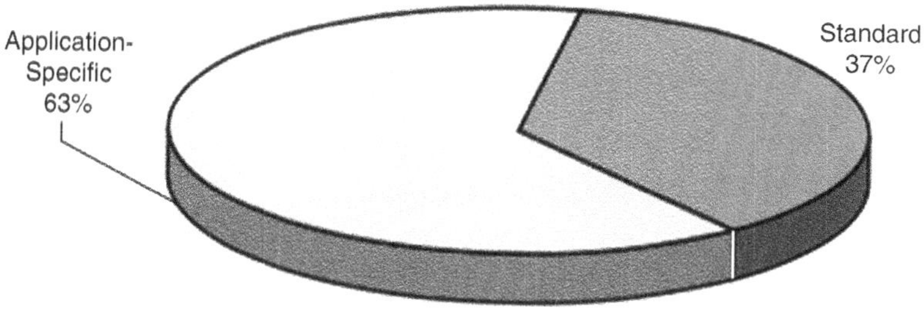

FIGURE 5-35 2006 analog IC sales by category (McClean et al., 2007).
Source: WSTS, IC Insights

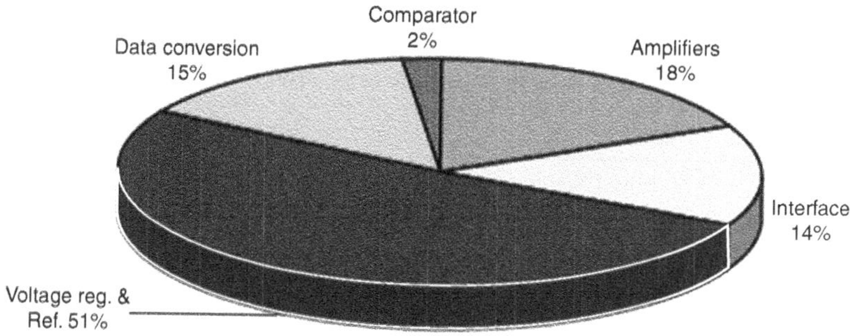

FIGURE 5-36 2006 standard analog IC market by category (McClean et al., 2007).
Source: WSTS, IC Insights

applications can also mean great business opportunities for analog IC manufacturers, as analog ICs are required to translate varying external phenomena, such as sound, heat, and pressure, into digital outputs for digital systems to manage and respond to them.

In Figure 5-36 and Figure 5-37 the analog IC market is shown by categories. New applications, such as smart phones, have prevented the decline of the analog IC market. According to Figure 5-37, communications (wired and wireless) contributed 48 percent of the 2006 application-specific analog IC market. Table 5-27 shows a sampling of applications.

Table 5-28 shows the top analog IC vendors for 2010 (Riemenschneider, 2010b; Anon., 2010h; Heise, 2010b). The top analog IC manufacturers share about half of the merchant market, leaving the rest for smaller manufacturers and niche vendors.

5.7.3 Application of Forecasting Methodology

Compared with microcomputers, memory, and microprocessors, analog parts do not have a unique evolutional parameter. Of course, their supply voltages and packages do evolve over time, but their key feature does not show evolutional trends. For instance, the capacity of memories, the key function

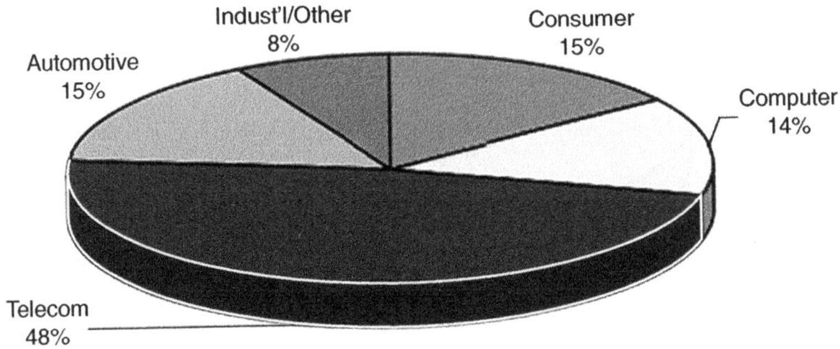

FIGURE 5-37 Application-specific analog IC market by category (McClean et al., 2007).
Source: WSTS, IC Insights

TABLE 5-27 Application of Analog ICs (Karls et al., 1998)

Analog IC Type	Manufacturer	Targeted Application
Op Amps	National Semiconductor	Portable electronics
	Texas Instruments (formerly Burr-Brown)	Video, medical imaging, signal processing
Data Converters	Analog Devices	Handheld instruments
	RF Micro Devices (formerly Micro Linear)	Smart phones, pagers
Voltage Regulators	Linear Technology	Portable CPUs, auto displays
	ST Ericsson	Battery-powered products

TABLE 5-28 2010 Analog IC Manufacturers Rankings (Riemenschneider, 2010b; Anon., 2010h; Heise, 2010b)

Analog IC Manufacturer	Sales in Billions of Dollars
Texas Instruments	15.7
STMicroelectronics	13.5
Infineon	8.2
NXP	5.3
Analog Devices	0.7
National Semiconductor	0.4

of all available memories, does evolve over time. Compared with that, the voltage regulation of a voltage regulator does not evolve at all.

As stated in the previous chapter, in order to be cost-effective, the procurement lifetime approach is chosen. Since this approach has already been performed in Chapter 4, there will be no detailed description in this section. For a detailed description of this approach, please refer to Section 4.2. The results of this approach can be found in the previous chapter (see Table 5-23). In order to compare the procurement lifetime results with the

Basic Life Cycle Methodology, a short assessment of the Basic Life Cycle Approach was performed, and the results are shown next.

Sales data of analog parts from the ICE (Karls et al., 1998) group were used to obtain least squares curve fits having Gaussian form.

5.7.4 Determining the Zone of Obsolescence

The primary attribute chosen for this analysis is part type. The secondary characteristics that may be applicable to various categories of part types and which should be used to modify life cycle estimates can include technology type, package style, and voltage.

Table 5-29 shows the values for the mean μ and standard deviation σ for the analog part types, together with the secondary attributes that should be considered for each category. Graphs of the life cycle models for these part types are shown in Figure 5-38.

TABLE 5-29 Mean μ and Standard Deviation σ Values for Various Analog IC Types

Analog IC Type	μ	σ	Secondary Attribute(s)
Amplifier	2003.4	8.3	Technology, package style
Comparator	2001.7	10.7	Voltage, package style
Voltage Regulator and Reference	2000.5	6.0	Technology, package style
Data Conversion	2009.6	11.9	Voltage, technology, package style
Interface	2001	8.1	Technology, package style

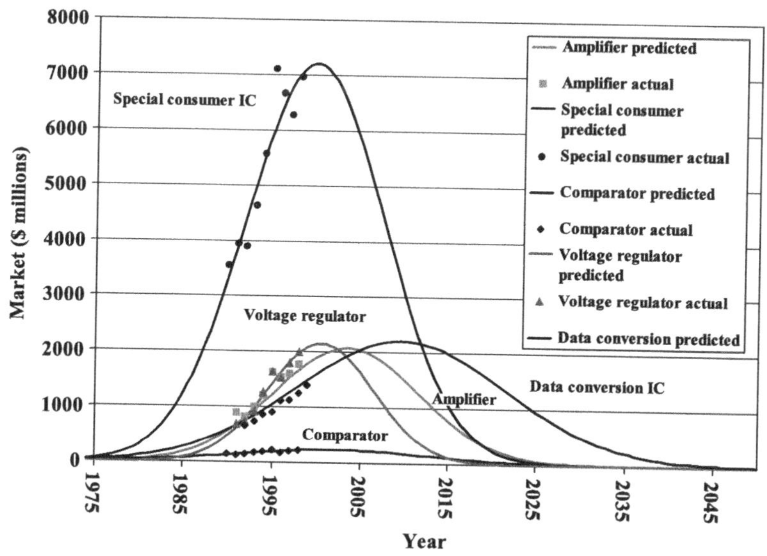

FIGURE 5-38 Life cycle curves for various analog part types.

To compare the forecasting results of the procurement lifetime forecast and the Basic Life Cycle Approach, the 200 mA low-noise RF low-dropout linear voltage regulator from ON Semiconductor was chosen.

According to the datasheet, this linear regulator is built for noise-sensitive RF applications such as power amplifiers in satellite radios, infotainment equipment, and precision instrumentation for automotive applications, which require very clean power supplies (ON Semiconductor, 2010). The NCV8570 is a 200 mA LDO that provides the engineer with very stable and accurate voltage with ultra-low noise and very high power supply rejection ratio (PSRR), which is suitable for RF applications. To optimize performance for battery-operated portable applications, the NCV8570 employs an advanced BiCMOS process to combine the benefits of low noise and superior dynamic performance of bipolar elements with very low ground current consumption at full loads offered by CMOS (ON Semiconductor, 2010).

The NCV8570 offers TSOP packaging and the following voltages: 1.8V, 2.5V, 2.75V, 2.8V, 3.0V, and 3.3V. For this comparison, the 1.8V option was selected. The Basic Life Cycle Methodology's results are shown in Table 5-30.

According to the Basic Life Cycle Approach, the linear regulator will become obsolete between June 2012 and March 2016. The procurement lifetime of a linear regulator, according to Table 5-23, is 11.63 years. The first revision of NCV7580 was released in May 2008. The latest revision is dated November 2009. Therefore, the procurement lifetime approach predicts the obsolescence in June 2020.

TABLE 5-30 Basic Life Cycle Analysis on NCV8570 from ON Semiconductor (p = 2010 [date of assessment])

Attribute	$\mu(\mu_j$ for modifying attributes)	$\sigma(\sigma_j$ for modifying attributes)	$(\mu + 2.5\sigma - p, \mu + 3.5\sigma - p)$	Zone of Obsolescence
Voltage Regulator	2000.5	6	(4.5, 11.5)	
BiCMOS	1997.2	3.2	(−3, −0.8)	Special Program
Package style (TSOP)	1999.7	5.2	(2.7, 7.9)	(2.7, 7.9)
Voltage (1.8V)	2003.6	3.6	(2.6, 6.2)	(2.6, 6.2)

5.7.5 Discussion of Analog Forecasts

At this point it is not possible to say which approach comes closer to the actual obsolescence date of NCV7580, but it is obvious that the procurement lifetime approach can be easily used for complex systems. Both methodologies can only be accurate when the latest complete data set was either available within a company or could be purchased from an outside market research firm.

For the primary characteristic—part type—the models developed in this analysis predict that analog ICs are now in decline. However, due to the relatively long life cycles of analog parts relative to many digital parts, a number of years remain before most of the parts will become obsolete. It is expected that over the next several decades, the stand-alone analog ICs that make up the discrete building blocks of current analog systems will gradually lose market share to integrated special purpose analog products or application-specific standard products (ASSP). Thus, discrete ICs such as amplifiers, data converters, references, interface ICs, and regulators will be replaced by integrated subsystems, such as telecom ICs, camcorder ICs, and health monitoring subsystems. System- or customer-specific analog ICs will be in greater demand as general-purpose products are deemphasized (Top Statistic Index, 2002; McAdam, 1999). Manufacturers are also increasingly integrating analog functions into what are generally considered digital ICs, such as DSPs. For example, TI offers the Stellaris ARM Cortex—M3 MCU (LM3S9B96), which includes 16 ADC channels (10-bit resolution) (Texas Instruments, 2010b).

As a customer-specific product approaches higher volume, analog and digital IC content can be further integrated into a cost-efficient, mixed-signal system chip. Document scanners are a good example of this integration. A discrete approach for a scanner requires 20 to 40 ICs, including the microcontroller and analog building blocks. An ASSP analog front-end approach, with associated microcontroller, digital ASIC, RAM, and motor-control ICs, decreases the chip count from 6 to 10 ICs. Finally, an integrated system chip design reduces the chip count to only three ICs, including RAM, leading to a low-cost, consumer-oriented scanner.

Analog parts will also see rapid changes in package styles in coming years, as predicted by the models. Advances in wafer bumping, together with increasing I/O count and performance requirements, will lead to the increased use of flip-chip technologies. Customer-specific packages (CSPs) will also be popular over the next several years. Since they can be assembled onto boards in much the same way as current surface mount components, while still taking up less space on the board, they represent a good transition technology on the path to flip chips (Adamson, 2000). Surface mount technology and convection-based reflow oven processing will remain dominant in the foreseeable future. Low-cost plastics will also continue to replace ceramic packages for most applications. Challenges for next-generation packages, which will increasingly house mixed-signal ICs, include decoupling, shielding, grounding, and testing (SIA, 1999).

The trend toward lower supply voltages is a result of the need for reduced power dissipation, reduced transistor channel length, and increased reliability of gate dielectrics (SIA, 1999). The advantages of low-voltage technology have been well understood for a long time; however, the difficulties with controlling threshold voltage and digital transition voltages within a narrow range have slowed their introduction. With improvements in process control techniques, it is now possible to reduce the operating voltage of a part without sacrificing quality or reliability. The drivers for the low-voltage market are mass-produced commercial electronics such as laptops, personal computers, and cellular phones. Soon, 3.0 volt and higher chips will be phased out and replaced by lower voltage parts. The International Technology Roadmap for Semiconductors sets the goal of reaching 0.3V by the year 2014 (SIA, 1999).

5.8 APPLICATION-SPECIFIC INTEGRATED CIRCUITS (ASICs)

This section presents an evolutionary approach to forecast the life cycle stage and the years to obsolescence of application-specific integrated circuits (ASICs). We then show how attributes such as package style and voltage can modify the forecasts. In addition, the trends in the ASIC industry, by ASIC category, are discussed and a road map is presented.

The term application-specific integrated circuit is a misnomer of sorts. An ASIC is actually customer-specific, rather than application-specific. For example, a gate array or standard cell IC is usually made for a customer. Programmable logic devices (PLDs) are considered to be ASICs because customers program them according to their needs (John and Smith, 1997).

ASIC design methodologies use integrated circuits containing arrays of prefabricated gates (gate arrays) or integrated circuits based on libraries of standard function cells (standard cell designs). These designs generally sacrifice density but allow the ASIC customers to perform significant in-house design, with help from manufacturers. Full custom IC designs address maximum density and performance for high-volume standard products.

Since the 1980s, the semiconductor industry has undergone substantial product changes. Memory parts, microprocessors, and other traditional high-volume standard products are referred to as commodity products. Current standard product portfolios include application-specific standard products (ASSPs), which differ from commodity products in that they incorporate value-added features for specific market segments (such as small computer standard interface [SCSI] controllers for workstations). ASICs differ from ASSPs in that they are customer-specific. Figure 5-39 shows the IC market hierarchy (Karls et al., 1998).

5.8.1 Types of ASICs

ASICs are subdivided into three main categories: custom ASICs, semi-custom ASICs, and programmable logic devices (PLDs). Each category is further divided into subtypes.

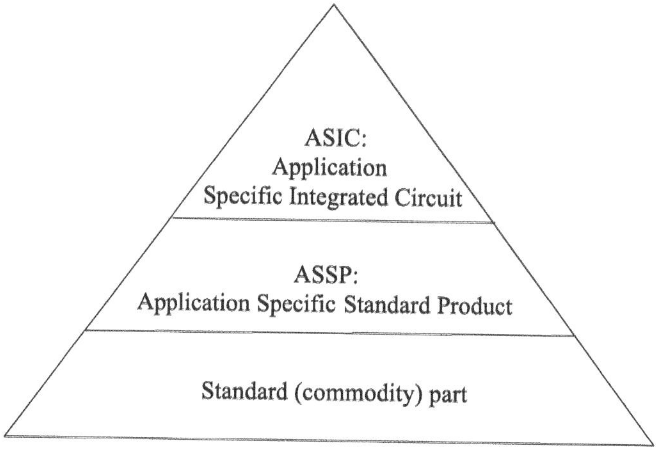

FIGURE 5-39 IC market hierarchy.

5.8.1.1 Full-Custom ASICs Full-custom ASICs are monolithic circuits, customized on all mask layers, with each cell optimized for a specific application. Although this design style was used exclusively for early ICs, engineers use it less now because of its high non-recurring engineering (NRE) costs and low design productivity (Harper et al., 1996). The exception is high-volume commodity products such as memory, which must be manually designed and optimized to meet density and performance requirements. Portions of high-end products, such as microprocessors, are often fully customized for performance.

5.8.1.2 Semi-Custom ASICs Semi-custom ASICs are monolithic circuits with one or more, but not all, customized mask layers. ASIC subtypes in this category include standard cells and gate arrays.

Standard cells are customized on all mask levels using pre-defined logic and function blocks made available to the designer in a cell library. Typical libraries begin with gate-level primitives such as AND, OR, NAND, NOR, XOR, inverters, flip-flops, and registers. Libraries contain complex functions such as adders, multiplexers, decoders, arithmetic logic units (ALUs), shifters, and memory (such as RAM, ROM). In some cases, the standard cell library may include complex functions such as multipliers, dividers, microcontrollers, microprocessors, and microprocessor support functions (parallel port, serial port, direct memory access, and event timers). Standard cell designs are generally created via synthesis from very high speed integrated circuit hardware descriptive language (VHDL). These standard cells can be purchased from certain IP (intellectual property) companies (such as ARM Ltd.). Standard cells are of two types: bipolar and MOS. CMOS standard cells are the fastest growing market segment. Leading standard cell ASIC manufacturers are shown in Table 5-31.

TABLE 5-31 Leading ASIC Manufacturers—Standard Cell, Gate Arrays, and PLDs (Karls et al., 1998)

Standard Cell ASICs
IBM Microelectronics
LSI Logic
NEC
Fujitsu
Gate Array ASICs
Fujitsu
Toshiba
NEC
Texas Instruments
LSI Logic
Programmable Logic Devices
Xilinx
Altera
Lattice
Microsemi
Cypress
Quicklogic
Atmel

Gate arrays are monolithic ICs that have columns and rows of transistors organized in blocks of gates. Compared with standard cell or full custom designs, the design turnaround time and cost are less. Wafer costs are low because different customer designs can be created from the same base wafer design. Foundry packaging and testing costs are also reduced due to standard pin-outs and common test fixtures that can be used for multiple designs (Harper et al., 1996). Gate arrays can be bipolar or MOS, depending on the fabrication technology. Linear arrays have transistors and resistors that perform the functions of linear ICs and discrete parts. Leading gate array ASIC manufacturers are shown in Table 5-31. The linear array market is also less popular.

5.8.1.3 Programmable Logic Devices Programmable logic devices (PLDs) are monolithic circuits with fuse, antifuse, or memory cell–based circuitry that may be configured (programmed), and in some cases reprogrammed, by the user in-system to perform a particular logic function or combination of functions. By eliminating the requirement to cycle through an integrated circuit production facility, the time to market can be reduced for a PLD. PLDs are gaining popularity because they provide low-cost ASIC solutions with nearly instant "manufacturing," as users can (re)configure them for their specialized needs. Programmable logic devices include, but are not limited to, programmable array logic (PAL), programmable logic array (PLA), field programmable gate arrays (FPGA), and erasable programmable logic devices (EPLD).

5.8.2 The ASIC Market

In general, designers are switching from full custom to standard cell designs. Among all ASIC types, the standard-cell approach is the most popular, due to the availability of advanced cell libraries including core analog, microprocessor, and special purpose blocks, which allow standard cells to be used in a broad range of applications. These advances are attracting high-end users who have traditionally relied on full custom designs. The catalyst for ASIC growth is MOS standard cell sales. The market success of the standard cell design reflects the increasing availability of core functions (Harper et al., 1996). Core functions simplify ASIC designs by providing entire subsystem or chip-level functional library cells from which the designer can fabricate the complete ASIC (Gifford, 1990).

The trend in the dynamic IC industry is toward high-performance and high-density technologies. Another PLD type, growing in popularity, is the MOS PLD, whose flexibility and quick time to market (two key issues for system designers) have helped the upward trend.

CMOS PLDs are in the growth and maturation phases. Leading PLD manufacturers are shown in Table 5-31. In 2010, Altera and Xilinx covered 80 percent of the PLD market (Arnold, 2010). Moreover, within Lattice Semiconductor and Microsemi, there are trends to transfer the concept of standard cells (IPs) into the analog world (Arnold, 2010).

5.8.3 Application of Life Cycle Forecasting Methodology

To forecast both the years to obsolescence and the life cycle stage, we use a model based on sales figures. A primary attribute determines the curve-fit of the sales data to the model, and modifying attributes translate the obsolescence prediction, depending on their degree of influence over the primary attribute. In the case of ASICs, the primary attribute is the ASIC type, and the dominant modifying attributes are package style and nominal supply voltage. We have used data from ICE (Karls et al., 1998) to develop the algorithm to forecast the life cycle stage and years to obsolescence of an ASIC part (Huang et al., 2001).

The procurement lifetime approach could be used for ASICs as well, but it requires data, including manufacturer's past product market introduction dates and end-of-life dates. This book focuses on the Life Cycle Forecasting Approach because the primary attribute, the ASIC type, shows clear evolutional trends.

Table 5-32 shows the values for the mean μ and standard deviation σ for the major ASIC types.

TABLE 5-32 Mean μ and Standard Deviation σ Values for ASIC Subtypes

ASIC Type	μ	σ	$\mu + 2.5\sigma$	$\mu + 3.5\sigma$
Bipolar Gate Array	1986.4	6.9	2003.6	2010.4
MOS Gate Array	1996.6	4.7	2008.5	2013.2
Linear Array	2009.2	16.2	2049.7	2065.9
Bipolar PLD	1989.1	3.7	1998.3	2002.0
MOS PLD	2000.5	4.9	2013.0	2017.9
TTL	1993.0	3.3	2001.4	2004.7
Bipolar Standard Cell	1994.5	3.7	2003.9	2007.6
MOS Standard Cell	2002.7	5.5	2016.4	2021.9
Full Custom	2000.7	11.5	2029.6	2041.1

5.8.4 Discussion of ASIC Forecasts

The value of the life cycle forecasting methodology has been demonstrated for off-the-shelf parts such as DRAMs, SRAMs, logic, microprocessors, and so on. ASICs differ from off-the-shelf parts in several ways; however, the critical differentiator from an obsolescence viewpoint is that for ASICs the customer has some ownership over the design information associated with a part. Because of this ownership, the interpretation of the sales curve is important.

With respect to ASICs, the sales curve could represent any one of the following conditions:

1. Ability or willingness to design the part. Loss of ability to design the part may be due to loss of the required design tool support, loss of human expertise, unwillingness to commit design expertise (too busy designing for newer technologies), or lack of availability of the appropriate intellectual property (IP)—IP obsolescence.
2. Ability or willingness to fabricate the part.
3. Lapsing of legal rights to use the necessary IP. The ASIC supplier has the technical capability but no longer has the necessary legal rights to design or fabricate a part.

For a DRAM that is purchased off-the-shelf, all that is relevant to the buyer is when the part becomes unavailable for purchase. On the other hand, if the sales curve represents a decrease in the willingness of ASIC suppliers to design a specific type of ASIC, but the buyer has complete ownership of the design (including all its IP), then the ASIC sales curve is not relevant in determining the obsolescence characteristics of the buyer's ASIC.

Proper interpretation of the sales curves is just as important when a buyer considers using an ASIC. In this case, if the projected sales curve is based on the ability to design, a buyer might make the decision to own more of the design than if the sales curves are interpreted as the ability to fabricate the part.

Another trend in ASICs is toward system-on-a-chip (SOC) development (Huber and Rosneck, 1991). With the advent of deep submicron process technology, easier access, knowledge of IP cores, and a market starting to demand integration, SOC designs have received a boost. Companies such as Texas Instruments have developed hybrid architecture—a TimeCell ASIC architecture that combines gate array technology with a standard cell. Texas

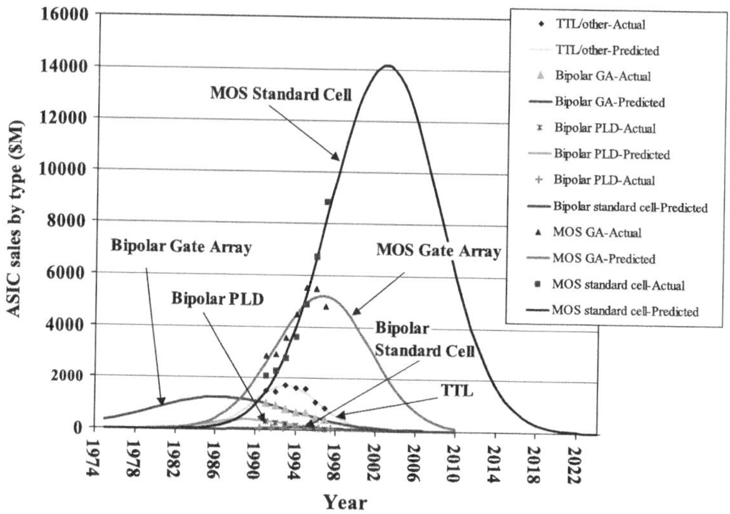

FIGURE 5-40 Life cycle curves for custom ASICs.

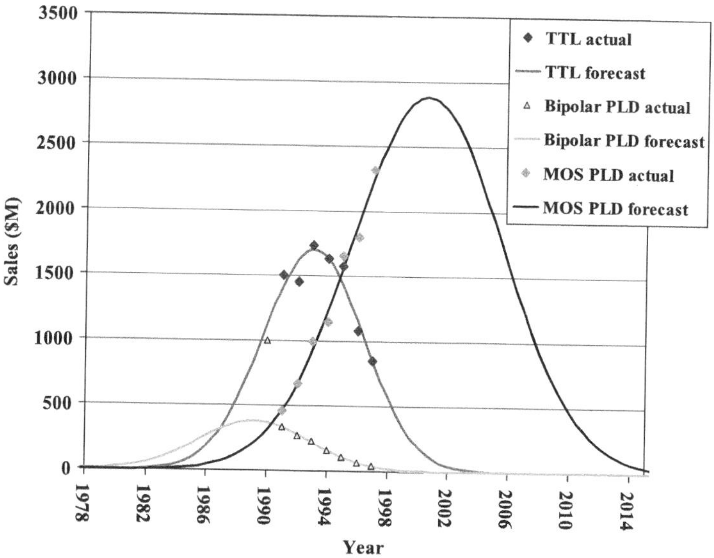

FIGURE 5-41 Life cycle curves of PLDs.

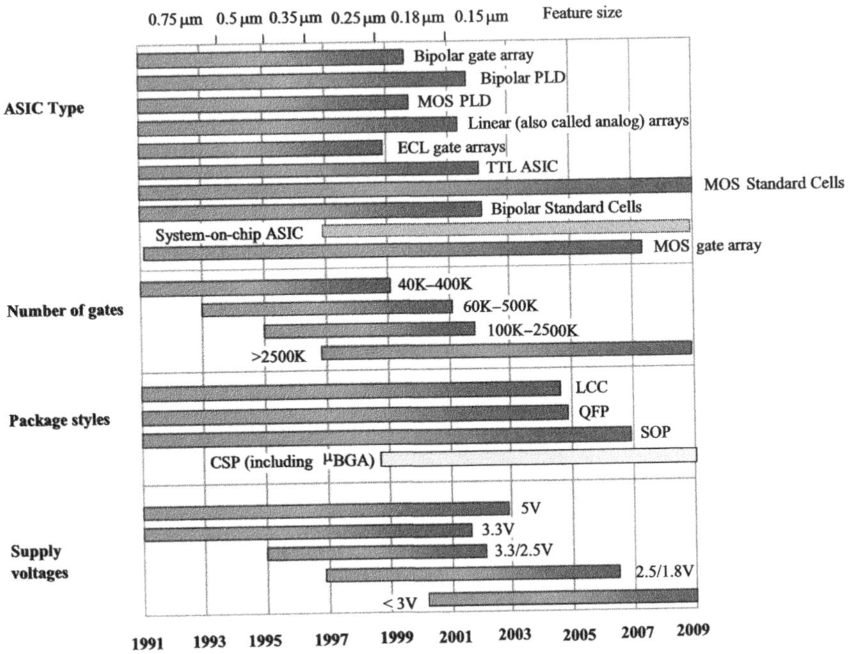

FIGURE 5-42 The history of ASICs.

Instruments is officially out of the gate array business with the introduction of TimeCell and is only going to invest R&D resources in future versions of standard cells and TimeCell ASICs.

Figure 5-40 shows the life cycle curves for the custom ASIC category. It shows that the mainstream IC market has moved away from bipolar-based ICs toward MOS-based technology, and full custom ASICs are continuing to lose market share to ICs such as standard cells.

PLD life curves are shown in Figure 5-41. Complex PLDs and FPGAs are expected to contribute to most of the growth in the PLD sector (Schroeter, 1992).

In summary, the MOS standard cell is forecast to have the fastest growth rate of all the ASIC categories. Figure 5-42 shows the history of ASICs.

Chapter **6**

Software Obsolescence

Today there are a growing number of methodologies, databases, and tools that address status, forecasting, risk, mitigation, and management of electronic parts obsolescence (Sandborn et al., 2003). However, the one common attribute of all the methodologies, databases, and tools that are available today is that they focus almost exclusively on the hardware life cycle. In most complex systems, software life cycle costs (redesign, re-hosting, and requalification) contribute as much or more to the total life cycle cost as does the hardware, and the hardware and software must be concurrently sustained.

Software obsolescence (more specifically, COTS, commercial off-the-shelf software obsolescence) is generally due to the following causes:

1. Technological Obsolescence[1]: The sales and support for the software terminate.
   - The original supplier no longer sells the software as new (end-of-sale)
   - The inability to expand or renew licensing agreements (legally unprocurable)

[1]Technological obsolescence is usually the result of more technologically advanced products introduced into the market, resulting in a phase out of the predecessor products and ultimately their end of support. This term has been used in the literature in the context of hardware components to mean the original equipment manufacturers stop manufacturing the component; however, in the context of software it means the original software developers no longer sell legal copies of the software, provide technical support, create updates or patches, or provide expansion packs to integrate with other software programs.

Strategies to the Prediction, Mitigation and Management of Product Obsolescence, First Edition. Bjoern Bartels, Ulrich Ermel, Michael Pecht and Peter Sandborn.

- Software maintenance terminates—the original supplier or third parties no longer support the software (end-of-support)

2. Functional Obsolescence: The changes in either hardware or software result in an incompatibility.
   - Changes in minimum hardware requirements cause the existing hardware to be obsolete (software-driven hardware obsolescence)
   - Changes in hardware result in an incompatibility with the software (hardware-driven software obsolescence)
3. Logistical Obsolescence: The ability to access the software is limited or terminates.
   - The digital media or network download used to distribute the software is no longer supported
   - The equipment and software used to build, test, and integrate the software is no longer supported

Analogously, hardware obsolescence can be categorized similarly to software obsolescence: technological obsolescence for hardware means that more technologically advanced hardware is available; functional obsolescence in hardware is driven by software upgrades that will not execute correctly on the hardware; and logistical obsolescence means that you can no longer procure a part.

The applicable definition of software obsolescence varies depending on the system that uses the software and where and how that system is being used. COTS software has both end-of-sale dates and end-of-support dates that can be separated by long periods of time. For many mainstream COTS software applications (such as PC operating systems), both the end-of-sale and end-of-support dates may be published by the software vendors. For applications that have a connection to the public web (servers and communications systems), the relevant software obsolescence date for both the deployment of new systems and the continued use of fielded system is often the end-of-support date because that is the date on which security patches for the software terminate, making continued use of the software a security risk. For other embedded or isolated applications, the relevant software obsolescence date is governed by either an inability to obtain the necessary licenses to continue using it or changes to the system that embeds it (functional obsolescence issues).

Although some proactive measures can be taken to reduce the obsolescence mitigation footprint of software, including making code more portable, using open-source software, and third-party escrow (a third-party receives source code if the original software developer terminates support) where possible, these measures fall short of solving the problem because they often require either the cooperation of the original software developer or large or simply unavailable resources. It is also not practical to think that software obsolescence can somehow be avoided. Just like hardware, safety, mission, and infrastructure critical systems have little or no control over the supply chain for COTS software or the software development infrastructure they may depend upon for developing and supporting in-house software.

In the COTS world, hardware and software have developed a symbiotic supply-chain relationship where hardware improvements drive software developers to obsolete software, which in turn causes older hardware to become obsolete. From Dell and Microsoft's viewpoint, this is a win-win strategy; consider the following quote from Bill Gates, the founder of Microsoft (Gates, 2003): "The only big companies that succeed will be those that obsolete their own products before someone else does." Besides COTS software (hardware specific and non-hardware specific), system sustainment depends on in-house (organic) application software, software that provides infrastructure for hardware and software development and testing, and software that exists at the interfaces between system components (enabling interoperability). Although hardware obsolescence precipitated software obsolescence is becoming primarily an exercise in finding new COTS software (more often COTS software and new hardware are bundled together), the most challenging software obsolescence management problem is often found at the interfaces among applications, applications and the operating system, and drivers.

Little attention has been paid to software obsolescence. Besides digital preservation,[2] the termination of sales and support of software comprises the primary focus of the existing literature (Rickman and Singh, 2002; Merola, 2006; Mittelstaedt, 1999; Sheppard, 2005; Gowland, 2002; CIE 2006).

6.1 THE ROOT CAUSES OF SOFTWARE OBSOLESCENCE

To enable the forecasting and management of software obsolescence, the root causes of the obsolescence need to be identified. Software obsolescence is specific to how the software is being used within a system. Figure 6-1 shows the root causes of software obsolescence. The three general causes of software obsolescence are categorized as technological, functional, and logistical and will be referred to as obsolescence modes. The next hierarchal-level down groups similar root causes of software obsolescence into what are referred to here as mechanisms. The purchase obsolescence mechanism includes obsolescence root causes that pertain primarily to purchasing issues such as the inability to purchase a legal copy of the software or the problem of not getting extensions to software licenses. The support mechanism has to do with obsolescence root causes that arise when the technical support of software ends or when updates or patches are no longer created for a software application. The compatibility mechanism deals with problems that develop when changes in either software or hardware result in an incompatibility between the two, eventually limiting or terminating the functionality of the system. The infrastructure mechanism

[2] If, for example, one searches on "software obsolescence" in Google, the vast majority of what appears is related to "information or digital preservation."

includes problems that arise when the tools needed to build, test, and integrate the software with other software applications become unavailable. The distribution mechanism encompasses problems that arise when the access to the software is limited or terminated, such as the inability to read the digital media on which the software is delivered because the interface (software or hardware) used to read the software is unavailable for various reasons.

Figure 6-1 shows the attempt to hierarchically model most obsolescence root causes. This model does not cover all possible causes of software obsolescence; rather, it is an effort to capture the predominate factors that contribute to software obsolescence.

Each obsolescence mechanism is a way in which software can go obsolete. The mechanisms are modeled as independent elements, meaning the metrics used to forecast the obsolescence event caused by one mechanism do not affect the metrics used to forecast another.

Not all five obsolescence mechanisms presented in Figure 6-1 will be applicable elements in the forecast of software obsolescence. Establishing which of the mechanisms should be included in the forecasting of a software's obsolescence is done by examining each of the five obsolescence mechanisms and asking the question "If the event this mechanism represents were to happen, would the system where the software is a component cease to complete its intended purpose and meet its design requirements?" If the answer is yes, then the mechanism should be included in the software's obsolescence forecasting; otherwise, the mechanism should not be included.

After determining which software obsolescence root causes and mechanisms are relevant in forecasting the software obsolescence, a lifetime for each of the relevant obsolescence mechanisms can be forecasted. A lifetime is a period of time whose beginning is marked by the introduction of the software to market and whose ending is marked by the obsolescence event of the software, which is defined by the mechanism that caused the obsolescence event to occur. The obsolescence event that occurs the earliest will be designated as the software's obsolescence date.

The next section will discuss the various methods used to forecast the lifetimes for each obsolescence mechanism identified in Figure 6-1.

6.2 SOFTWARE OBSOLESCENCE MECHANISMS

The key enabler for proactive and strategic management of DMSMS obsolescence is the ability to forecast the obsolescence events for key hardware and software elements in systems.

Most long-term obsolescence forecasting for hardware (electronic parts) is based on the development of models for the part's life cycle. Traditional methods of life cycle forecasting are ordinal scale–based approaches, in which the life cycle stage of the part is determined from a combination of technological and supply chain attributes, such as level of integration, minimum

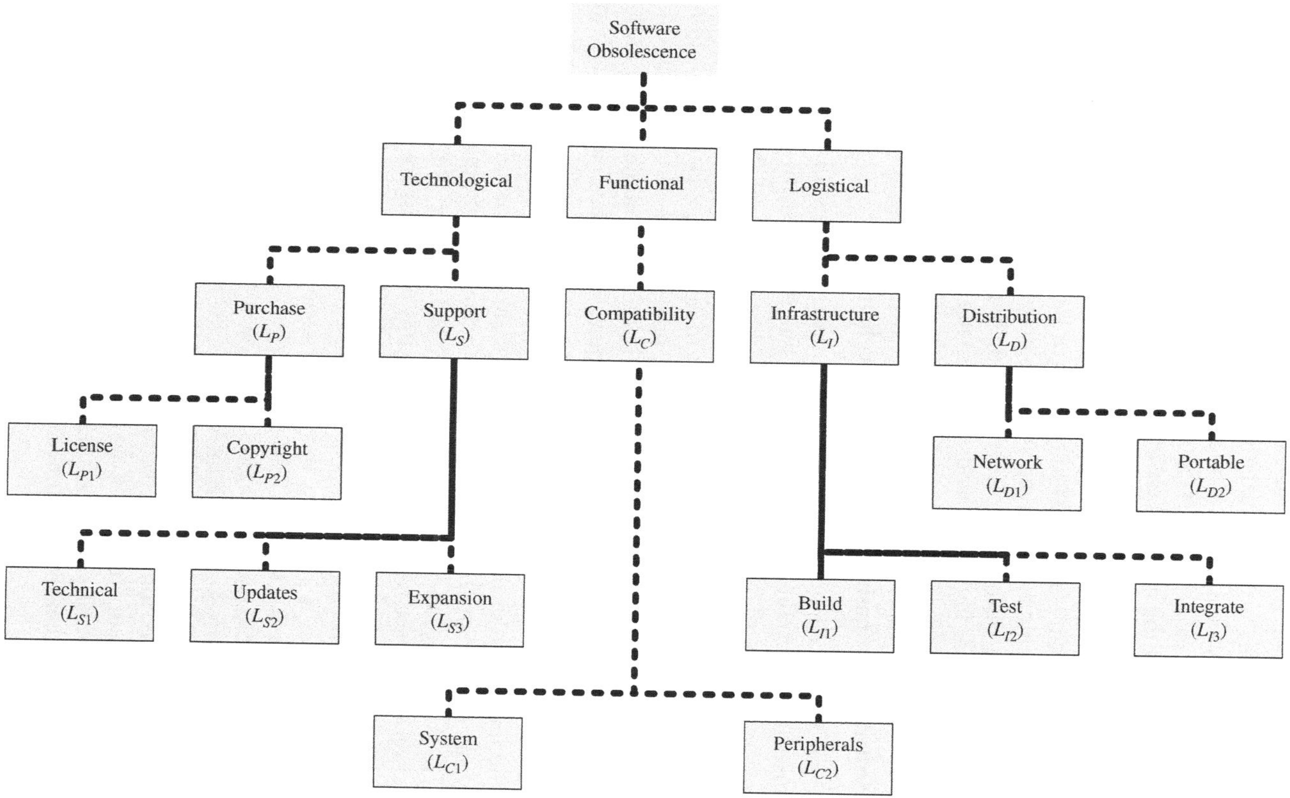

FIGURE 6-1 Software Obsolescence Modes, Mechanisms, and Root Causes.

feature size, type of process, number of sources, and so on (Henke and Lai, 1997; Josias and Terpenny, 2004), and those available in several commercial databases. Market data has been used to forecast life cycle curves, from which obsolescence forecasts are obtained (Solomon et al., 2000; 2007), and data mining–based obsolescence forecasting approaches that leverage large commercial databases of electronic parts have also been developed (Sandborn et al., 2007; 2011). None of the methods developed for forecasting hardware obsolescence have been applied to software.

Forecasting software obsolescence will involve finding a metric that is directly related or highly correlated to the root cause of the software obsolesce. The methodology behind forecasting software obsolescence for each mode of obsolescence will be different; however, all methods in general will involve gathering metric data, finding a trend, and deriving a relationship.

6.2.1 Software Purchasing Obsolescence Mechanism

The software purchasing obsolescence mechanism includes all obsolescence root causes that are due to problems with purchasing software. The two primary root causes of this mechanism are the inability to purchase legal copies of software and the inability to purchase extensions to existing software licenses. The software purchasing lifetime (L_P) is the duration from market introduction to the end of sale (end of sale of legal licenses). Determining the software's purchasing lifetime can be done analogously to the procurement lifetime of electronic parts (hardware) developed in Sandborn et al. (2011).

Unlike hardware, in most cases the introduction of the next software version means the immediate discontinuance of sale of the previous version (there is virtually no overlap in the availability for purchase of multiple versions of software). Analyzing the time between the last update and the market introduction of the next version demonstrates this trend (Figure 6-2).

The software market works with licenses, so discontinuance of the software means that no more licenses will be sold. Software developers often provide the right to downgrade if a license of the newer version is purchased. Since this downgrading right and extended OEM license procurement are highly software developer specific (Microsoft gives two years downgrading right and continued two years OEM licenses after the introduction of the successor version), the introduction date of the successor software will be interpreted as the end-of-life date of the former version, while the introduction of the latest software might obsolete its predecessor (from a software purchase viewpoint), but does not in general mean the end of support for its predecessor.

The Software Purchasing Lifetime (L_P) is defined as

$$L_P = D_{EOL} - D_{PUB} \tag{6-1}$$

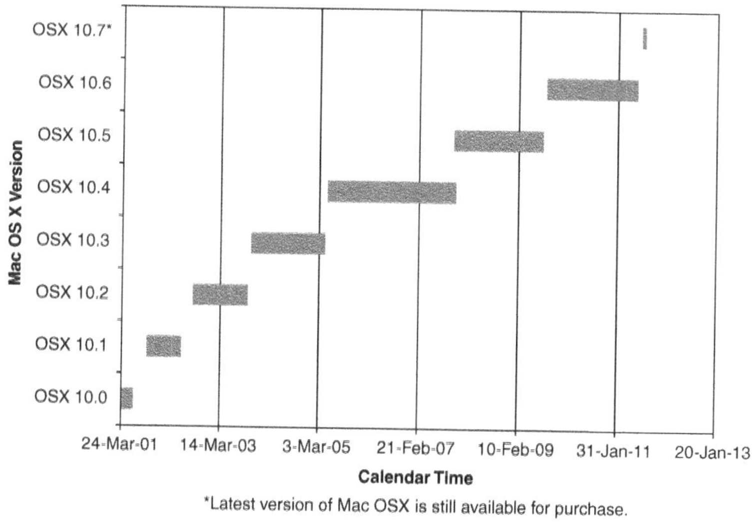

FIGURE 6-2 Calendar time between the last update and the market introduction for each version of Mac OS X.

where

D_{PUB} = software initial release date: the date software is first purchasable.

D_{EOL} = end of life date: the date the software developer stops selling new licenses to this software. This generally corresponds to the introduction date of the software's successor version.

6.2.2 Software Support Obsolescence Mechanism

The software support lifetime is the duration from market introduction to the end of support (end of technical support) for the software. In comparison to electronic parts (hardware) lifetime or procurement lifetime, software obsolescence is not only defined by the software developer's discontinuance date, that is, the end of sales does not implicate its end of support. For example, the official sales stop of Microsoft Windows XP was on June 30, 2008, but the mainstream support stopped on April 14, 2009 and the extended support stops on August 8, 2014. The extended support includes monthly security updates that enable the continued safe usage of Windows XP. By purchasing an early license of Windows Vista, downgrading to Windows XP Professional was made possible, and special OEM exceptions were provided for ultra-low-cost PCs until June 30, 2010. This example shows that the sales stop of software does not mean the end of its support lifetime. The software support lifetime (L_S) is defined in Equation 6-2:

$$L_S = D_{EOS} - D_{PUB} \tag{6-2}$$

where

D_{PUB} = software initial release date: the date software is first purchasable.
D_{EOS} = end of support date: the date software updates cease and support is no longer available from the software developer.

Compared with electronic part discontinuance notices, software end of support dates are often published by the software developer. For instance, Microsoft publishes end of support dates for most of its software products on the Internet. This makes software support a candidate for a data mining–based approach to forecasting its obsolescence. To demonstrate a data mining–based approach for software support lifetime, software versions based on Microsoft Windows NT were analyzed.

Figure 6-3 is a plot of software support lifetime versus introduction date for all Microsoft Windows versions based on Windows NT. The graph in Figure 6-3 only contains the official end-of-support information.

Using a best fit linear approximation, the support lifetime of a future NT-based version in years can be forecasted as

$$L_s = 0.2168(D_{PUB}) - 424.45 \tag{6-3}$$

where D_{PUB} is a date.

Equation 6-3 is software developer and software specific. In general, software developers are focused on the introduction of new products rather than the support of existing products with decreasing sales.

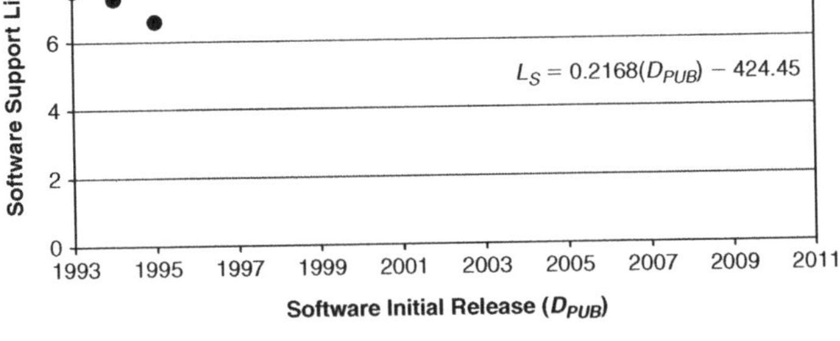

FIGURE 6-3 Software Support Lifetime versus introduction date for Microsoft Windows NT Software.

6.2.3 Software Compatibility Obsolescence Mechanism

Functional lifetime is the duration from market introduction to when software requirements exceed the existing system operational hardware or the existing non-system operational hardware (such as test systems). Functional software obsolescence occurs if there is a time gap between the evolution of software and hardware.

The minimum system requirements for software will eventually exceed existing hardware resources. Moreover, improved hardware might complicate the usage of existing software (Mac OS X 10.4 might not run on Intel-based computers). Nevertheless, the most common problem is that software requirements exceed the existing system's hardware resources. To exemplify this effect, compare system requirements of Microsoft Windows XP and Microsoft Windows 7 (Figure 6-4).

$$L_C = D_{MRHC} - D_{PUB} \tag{6-4}$$

where

D_{PUB} = software initial release date: the date software is first purchasable.

D_{MRHC} = minimum hardware requirement change date: the date when the next version of the software becomes available for purchase that has a change in the minimum required hardware as compared to the previous version (disk space, RAM capacity, processor speed and type).

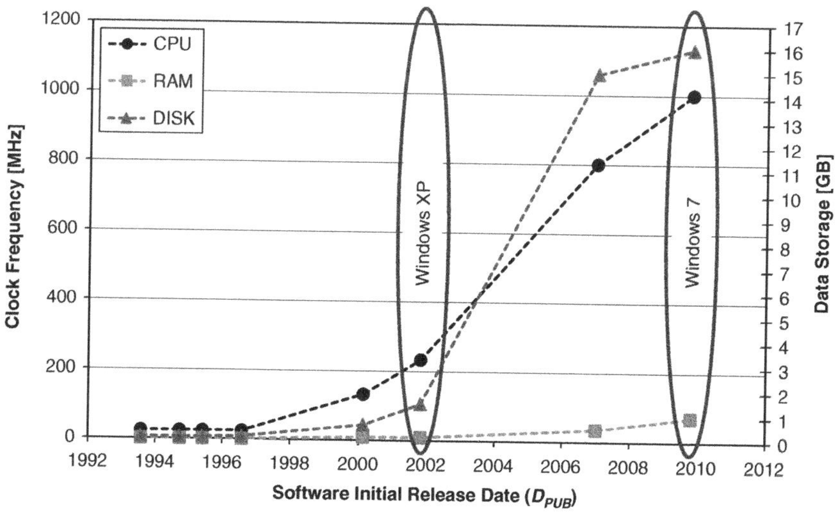

FIGURE 6-4 Evolution of minimum system requirements for Microsoft NT-based Windows.

According to the latest trends in computing, there is a possibility that the coupling between hardware and software on the end user's computer may be shrinking. Therefore the calculation for this life cycle time might become more difficult. Cloud computing allows anyone with Internet access to utilize the resources of server computers. In this case, the software functional life cycle time then applies for server computers that form the basis for cloud computing. Software's minimum hardware system requirements might be specific to particular kinds of software. For example, engineering and simulation software might not require the same hardware resources as graphical editing software. Thus, forecasting the functional life cycle time on a general basis is complicated. In order to achieve the best forecasting accuracy, the software specific system requirements must be considered. To demonstrate a forecasting approach to functional life cycle time, Microsoft Windows XP and Windows 7 were chosen. Figure 6-4 shows the evolution in the minimum system requirements to run each version of Windows NT. Two of the most popular versions of the Window NT series are Windows XP and Windows 7, which were first introduced in October 2001 and October 2009, respectively. Looking at those two versions of Windows NT as indicated on Figure 6-4, it is evident that within a decade the minimum system requirements have increased dramatically.

To run Windows 7, a four times faster processor, 15 times more RAM, and about 10 times more hard disk space is required, compared to Windows XP. As far as the graphics are concerned, significant improvements lead to new standards like Direct X9 and Windows Display Driver Model, but is this hardware requirement increase an accurate indicator for future operating systems? See Figure 6-5.

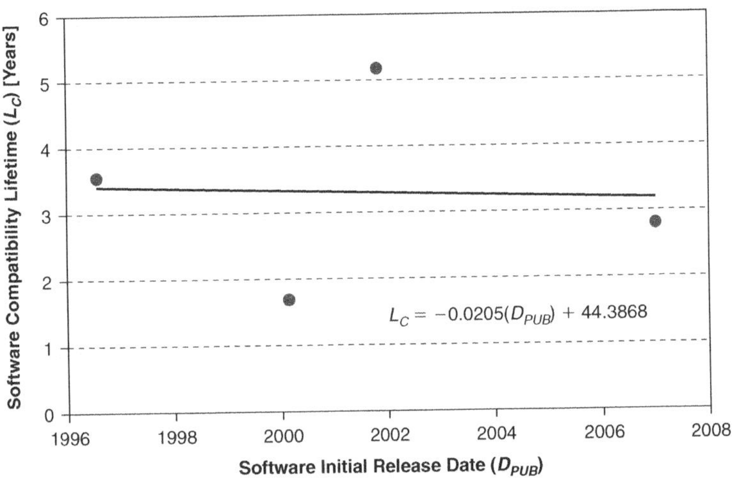

FIGURE 6-5 Time between changes in minimum required hardware.

Using the following trend equations, the Microsoft Windows Release's minimum requirements in October 2017 can be forecasted using

$$\textit{Minimum CPU clock frequency}\ [\text{MHz}] = -4.7595653(D_{PUB})^2 + 19{,}189.56349(D_{PUB}) - 19{,}340{,}891.53$$

$$\textit{Minimum RAM}\ [GB] = 0.013253215(D_{PUB})^2 - 53.057438(D_{PUB}) + 53{,}102.18$$

$$\textit{Minimum DISK}\ [GB] = 0.356445(D_{PUB}) - 700.3$$

The trend equations above are valid for $D_{PUB} > 2001$ (for CPU and RAM) and > 2007 (for DISK). Thus the clock speed is forecasted 1.16 GHz, the hard disk space 19 GB, and about 3.6 GB memory.

It also has to be mentioned that hardware platforms might exceed the software's needs. For example, in January 2010, Windows expert Dirk Makowski performed an experiment running Windows NT 3.1 on an Intel Core2Duo e6600 (two cores with each 2.4 GHz 64-bit, 6GB RAM, and 640GB hard drive) (Makowski, 2010b). From the results, he stated that there were significant changes in interfaces and driver support (using SATA hard drives is not possible). As far as I/O is concerned, no USB mouse can be used; it is required to adapt to PS2. Due to the advanced 64-bit processors, there is a lack in compatibility with the former i386 structure. This leads to the fact that Windows was unable to boot. After various conservative settings in the BIOS, the DOS environment and Windows NT 3.1 standard mode could be accessed by DOS. In this mode, the operating system behaves like a fast Motorola 80286 system. Nevertheless, the system could no longer be used for various applications, due to occurrence of division by zero errors while executing programs. This error leads to abnormal program ends, since the old algorithms are calculated too fast and finally result in severe timing problems (Makowski, 2010b). To summarize, there were an array of hardware-software incompatibility problems such as processor structure, drivers, interfaces, and software algorithms.

6.2.4 Software Infrastructure Obsolescence Mechanism

The infrastructure mechanism obsoletes software by making the tools needed to build, test, and integrate the software with other software applications unavailable. For example, the computer program in some systems are written in assembly language, which requires a compiler to translate high-level computer code into low-level code. Of these systems many require a compiler that can no longer run on any computer but require specialized workstations that perform this specific operation; however, since these workstations are no longer being produced, there is a possibility that when a system needs to be reprogrammed, the workstation needed to compile its guidance or targeting program is no longer available.

6.2.5 Software Distribution Obsolescence Mechanism

Similar to the infrastructure mechanism, the distribution mechanism obsoletes the software by making the thing that can access the software from its original digital medium (DVD, CD, floppy disc, and so on).

Software technical life cycle is the duration from market introduction to when the media that the software is distributed on is no longer supported. Referring to Makowski's experiment, Windows NT 3.1 on Intel Core2Duo e6600 described in Section 6.2.4, the Windows NT 3.1 installer software was only available on 3.5-in floppy disks. In order to understand the technical life cycle time of software, former Microsoft Windows versions have been analyzed based on their delivery media (Table 6-1).

Forecasting software's technical life cycle times requires a detailed analysis of the common media market. This market is highly evolutionarily driven and therefore the forecasting methodology in Chapter 4 (Section 4.1) could be applied.

Today, 3.5-in floppy disk drives with USB connection are available, but there is no such interface modification to 5.25-in floppy disk drives available. The CD-ROM used to be a common data carrier for about 10 to 15 years and is slowly being superseded by DVD and direct downloads.

Increasing downloading capabilities will simplify the delivery media issue in the future. Nevertheless, these increased capabilities will also lead to future software requiring more data space.

The best way to eliminate this obsolescence factor is to re-record the software data to the latest data carriers. More details are discussed in the strategic obsolescence management chapter (Chapter 9) with the product specific Obsolescence Management Plan.

TABLE 6-1 Microsoft Windows Versions and Provided Delivery Media (Makowski, 2010a)

Version	Windows 3.0	Windows NT 3.1	Windows 95	Windows 2000	Windows XP	Windows 7
Data carriers	5.25" and 3.5" disks	3.5" disks and CDROM	3.5" disks and CDROM	CDROM	CDROM	DVD Download Optional CDROM

6.3 DISCUSSION

So where do we go from here? COTS software obsolescence is neither well understood nor managed today, and the specific impacts on system software are largely ignored throughout the existing DMSMS tool/database set. In reality, DMSMS is a hardware/software co-sustainment problem, not just a hardware sustainment problem. In existing tools, systems are treated as disembodied BOMs with little or no coupling from part to part and little functional knowledge of the operation of the system. Software obsolescence (and its connection to hardware obsolescence) is not well defined, and current DMSMS planning tools are not generally capable of capturing the connection between hardware and software.

Few, if any, system development and support organizations actually track and manage software obsolescence. Systems engineering approaches to concurrently manage software and hardware obsolescence are virtually nonexistent, and there are no formal organizations sharing obsolescence data and information on software. Various DMSMS working groups throughout industry and the defense community have only just started to address the management of software obsolescence impacts and welcome discussion on this topic.

Constraints imposed on the management of software obsolescence are also not well understood. The functional dependency between software and hardware and its effect on obsolescence management constraints further complicates this matter. The obsolescence of either software or hardware can force a design refresh that will design both components out of the system simply because one component will not work without the other. Based on this effect, another prominent issue is the mismatch between the revision cycle of software and the procurement lifetime of hardware. Whether a piece of software is getting an update or going through a revision to the next version, if the hardware's functionality is dependent on a certain software version or configuration, then any change to the software can potentially force the hardware to be obsolete. Since the time between major changes of software can be as fast as or even faster than the procurement lifetime of the hardware, this potentially can lead to a high number of design refreshes.

Chapter 7

Reactive Obsolescence Management

This chapter discusses the reactive management approaches that can be employed to mitigate obsolescence. These tactics and strategies can be short-term solutions adopted for a particular case to bridge to a future design refresh or longer-term efforts employed until the end of support of the system. Depending on factors such as system support life, period, and volume of production, the type of part, and plans for future design refreshes, the equipment manufacturer must decide which strategy to use to best handle part obsolescence issues when they occur (Wilkinson et al., 1999; Condra, et al., 2000; Porter, 1998a; Condra, 1999; Wasson and Wilkinson, 2001; Cavill, 2002).

The reactive obsolescence management approach consists of determining an appropriate resolution to the unavailability of a component that is already obsolete or the impending unavailability of a component that will become obsolete in the near future, executing the resolution process, and documenting and tracking the actions taken (Condra et al., 1997; Condra and Matthews, 1999).

Many tactics and strategies can be employed to deal with obsolescence problems. These are described in detail in the sections that follow. At the end of this chapter there is a reactive obsolescence management checklist that includes the factors that need to be considered when choosing the most appropriate mitigation approach. Furthermore, reactive obsolescence management guidelines are provided at the end of this chapter that describe how to implement a reactive obsolescence management system within an organization.

Strategies to the Prediction, Mitigation and Management of Product Obsolescence, First Edition. Bjoern Bartels, Ulrich Ermel, Michael Pecht and Peter Sandborn.

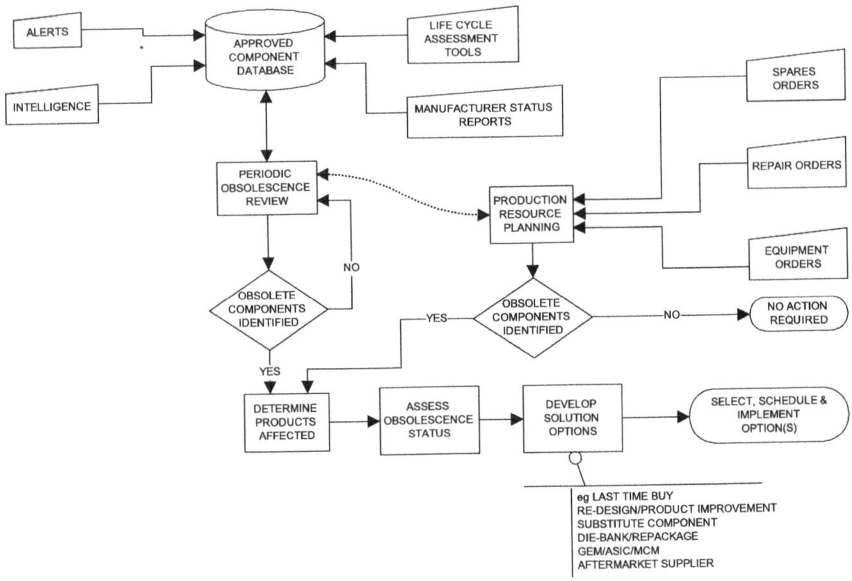

FIGURE 7-1 Smiths Industries obsolescence management plan (Wilkinson, et al., 1999).

Most reactive obsolescence management organizations develop detailed plans for how obsolescence events will be resolved when they occur using reactive approaches. There are also a number of publicly available guidance documents that can assist in preparing an organization's plan. These are also described in the sections that follow.

Figure 7-1 shows a sample obsolescence management plan developed by Smiths Industries, an aerospace supplier. This plan relies on obtaining information regarding obsolescence and then reacting to this information. Similar plans are used within the various defense programs in the United States and Europe.

The development and implementation of a plan such as that shown in Figure 7-1 is only the first stage in structuring an organization to manage obsolescence. Organizations that manage obsolescence successfully do so by recognizing the impact it has on the way systems are designed and supported.

7.1 CHANGE AND DISCONTINUANCE NOTIFICATIONS

A typical time line for product discontinuance is shown in Figure 7-2. The part manufacturer will issue a product discontinuance notice (PDN) or product change notice (PCN). The part user then has commonly three to six months in which to place final orders for the original part. The final orders may then be delivered over the following six months to one year. After this date, no more original parts are available from the original part manufacturer, although they may remain available from aftermarket suppliers in some cases.

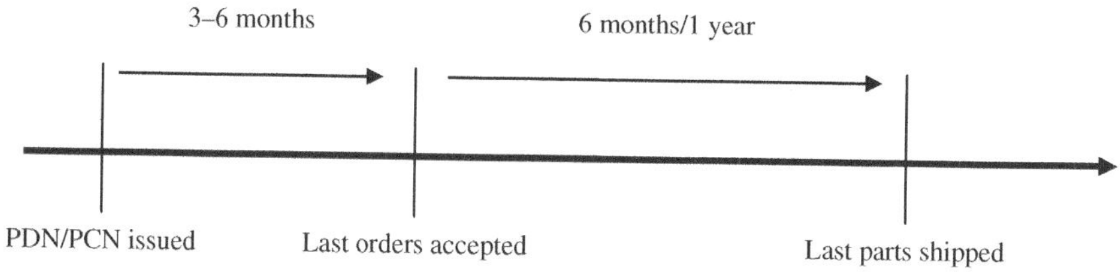

FIGURE 7-2 Example product discontinuance/change notice time line.

Component manufacturing companies issue a product discontinuance notice (PDN) when they decide to cease production of a part or replace it with a new part that is sufficiently different to warrant a new part number.

If component manufacturing companies change their products, they issue a product change notice (PCN) to inform their customers about the actual changes to parts.

The impact of a PDN is effectively the same as a PCN when the changes described in a PCN are major, making it impossible to replace the original part with the changed one. Especially in sensitive product sectors (for example, aerospace components), small changes to parts can make requalification of the part or the entire system in which the part is used necessary. The consequence is the effective loss of manufacturers or suppliers of items or raw materials, which is known as DMSMS, or obsolescence. In the following sections, PDNs and PCNs are discussed (Chapter 3 discusses different characteristics of changes and notification paths).

The timing and content of PDNs and PCNs are of vital importance to an equipment manufacturer, since in many cases the window of opportunity in which the equipment manufacturer has to take corrective action is defined by the contents of these notices. The key contents of such notices are the part numbers affected, the effective date of the notice, the last date on which orders for the part will be accepted, and the last date on which shipments of parts will be made. Sometimes part manufacturers will provide suggestions for replacement parts, but these are often not exact form-fit-function replacement parts and may require some re-engineering and requalification to successfully incorporate them. At the very least, the bill of materials, production drawings, and maintenance manuals have to be updated. The notice may also indicate whether the manufacturers have entered into any agreements with so-called "sunset suppliers" (or approved aftermarket suppliers) that take the remaining parts stock or transfer manufacturing information and tooling for continued manufacture.

Part manufacturers decide what to include in PDNs and PCNs and when to release them. There is currently no common standard for the information contained in PDNs and PCNs, nor any general legally binding legislation that requires their issuance (although individual customers may be able to insert PDN and PCN notification and content requirements into their specific contracts with manufacturers). The earlier that part manufacturers release the notices, the better for the equipment manufacturer. Some features of standardized end-of-life (EOL) notices and PCNs useful in managing obsolescence are listed below:

- Part number(s)
- Cross-reference of part number to cover industry and government standards such as National Stock Number (NSN[1]), JAN, and SMD
- Effective date of the notice
- Last purchase order date for affected parts
- Description of change
- Temperature grade (for electronic parts)
- Package type (for electronic parts)
- Product family and technology
- Possible alternative parts (if available)
- Continuing supply sources (if available)
- Lifetime (or last-time) buy (LTB) date
- Last-time delivery (LTD) date
- Warranty conditions associated with last-time buy parts
- Pricing information
- Availability and sourcing information of unpackaged dies (for electronic parts)
- Names and addresses of aftermarket and emulation sources (if applicable)
- Legal and business conditions

Once a PDN or PCN arrives at an equipment manufacturer, the process of finding an appropriate strategy and implementing suitable reactive tactics to combat and reduce the adverse effects of obsolescence begins. If the relevant PCNs and PDNs are not forwarded to the appropriate groups or individuals for assessment in a timely manner, lifetime buy opportunities can be missed.

7.2 OBSOLESCENCE RECOVERY (MITIGATION) TACTICS

There are a number of possible solutions to obsolescence problems after they occur. These range in complexity from a simple part substitution to a major redesign of a product. The selection of the most appropriate solution is a complex task involving a large number of factors, including the time available, the expected future production and support lifetime, and the expected occurrence of future product developments.

[1] A National Stock Number, or NATO Stock Number (NSN), is a 13-digit numeric code, identifying all standardized material items of supplies as they have been recognized by all NATO countries.

TABLE 7-1 Nonrecurring Resolution Cost and Time Metrics (FY2011) from the 2009 DMEA Survey (adapted from Shaw, et al., 2010)

Resolution type/ Strategy	Nonrecurring Resolution Cost (US$)	Section
Negotiation with the manufacturer	$0	7.2.1
Existing stock	$0	7.2.2
Reclamation	On average $20,000	7.2.3
Alternate part[2]	On average $41,000	7.2.4
Part substitution	On average $34,000 for a normal substitute($3,000 for an administrative substitute)($5,000 for a desktop substitute)($423,000 for a complex substitute)	7.2.5
Uprating	On average $34,000 (Uprating is a subset of part substitution)	7.2.6
Aftermarket sources	On average $33,000	7.2.7
Emulation[3]	On average $73,000	7.2.8
Redesign	On average $1,118,000 for a COTS part($1,094,00 for a custom part)($1,010,00 for a next higher assembly)	7.2.9
Reverse-engineering	This resolution is program specific and is dependent on individual DMSMS problems.	7.2.10
Lifetime Buys/bridge buys	This resolution is program specific and is dependent on individual DMSMS problems.	7.2.11

The cost of resolving diminishing manufacturing sources and material shortages (DMSMS) problems is of primary concern to program managers. Using data from many organizations, the DMEA, in conjunction with ARINC and CALCE, developed nonrecurring engineering cost factors for each resolution approach. The goal was to report the cost impacts of DMSMS problems and highlight the fact that the implementation of an effective obsolescence management system is necessary for companies and other organizations (McDermott et al., 1999; Shaerer and Tomczykowski, 2001; Shaw et al., 2010).

The various strategies available and their driving factors are discussed in the following sections, which are presented in approximate order of increasing resolution cost, as shown in Table 7-1 (Shaw et al., 2010).

The first report on resolution cost factors for DMSMS was published in 1999 (McDermott et al., 1999). In a supplemental report from 2001, these cost factors were validated by forecasting the nonrecurring engineering cost factors for each resolution and strategy for the years 2002 to 2006 (Shaerer and Tomczykowski, 2001). The current report and numbers are based on the 2009

[2]Alternate source includes parts from a different manufacturer (not already in the applicable technical data package) that meet the part specifications.

[3]Emulation cost values provided by Defense Logistics Agency (DLA) are not from the DMEA survey and represent the historical costs to the DLA to emulate a part.

DMEA DMSMS cost resolution survey results and were used to calculate new nonrecurring engineering (NRE) resolution cost metrics for the DoD DMSMS Guidebook of Best Practices (SD-22, 2010; Shaw et al., 2010).

7.2.1 Negotiating with the Manufacturer

A manufacturer's end-of-life (EOL) decision is rarely stopped or delayed. However, during the period between part phase-out and discontinuance, an equipment supplier may be able to persuade the manufacturer to keep the part from being discontinued by teaming with other end users (distributors) to prove a sufficient demand for the part or accept a price increase.

The manufacturer's decision may depend on its profitability and manufacturability objectives. If a sufficient volume of orders can be guaranteed, the manufacturer may be willing to continue production of the part. Several equipment suppliers may band together to share information (consolidate demand), develop common or standardized part lists, and provide avenues for communication with component manufacturers. Some manufacturers have responded positively when approached by a group, since enough demand is ensured to justify continued production of a part. In March 1997, Boeing encouraged their suppliers in a meeting to act cooperatively while dealing with semiconductor manufacturers. The collective equipment suppliers' strategy is to appear as one virtual customer to the IC makers (Pecht et al., 1997). This strategy is also applicable to all other product sectors beyond the IC market.

The advantages of negotiating with the manufacturer to keep a part active are that the original part is still available, no other system changes are required, and no nonrecurring engineering costs arise.

Suppliers may be hard-pressed to gain enough industry support and, if successful, may be forced to accept price increase and minimum volume commitments from the manufacturer.

Negotiation with manufacturers occurs in any form of business where a customer may want a part manufacturer to continue making a part that the part manufacturer wants to discontinue. The part manufacturer may continue manufacturing the part, may direct the customer to an aftermarket source, or may suggest substitute parts that they manufacture or are manufactured by another part manufacturer (Texas Instruments, 1999f).

7.2.2 Existing Stock

The use of existing stock describes searching for original parts that are available from stock that has been under the control of the equipment manufacturer since it was purchased directly from the original manufacturer (or distributor) (McDermott et al., 1999). To detect stock, the possible locations of these parts have to be evaluated. Analyses of an organization's own stock of inventory, supply depots, and all possible original suppliers and distributors should be included (DIN EN 62402, 2008). The number of available or purchasable parts should be enough to ensure supply during the whole remaining life cycle of the system.

If obsolete parts are still available from existing stock, this is an inexpensive resolution to an obsolescence problem, as the only costs that might arise are for functional testing of the parts and inventory. Testing is not only essential to prove the functionality, it is also necessary for protecting against counterfeit parts (see Section 7.2.7).

7.2.3 Reclamation

Reclamation (also referred to as harvesting) is the process of salvaging or "cannibalizing" used parts that have a remaining useful life.

The process of reclamation involves the following steps:

1. The parts that are required for production but are not available must be identified.
2. Possible sources from which the same or equivalent parts can be reclaimed must be identified. The possible sources of reclaimable parts are:
   - Systems beyond "economic repair"
   - Spares stocked for discontinued systems
   - Excess available stocks
   - Retired assets

Reclamation is a solution for obtaining small numbers of obsolete parts. It is also an in-house program with no dependency on outside agencies and therefore minimal exposure to counterfeit part risks.

Reclamation, however, involves major handling and rework efforts, logistics, and disassembly. Quality and reliability assessments of reclaimed parts should be conducted, though this may prove difficult as there is usually a lack of history attached to these parts. This poses a risk to the producibility and reliability of a product or system.

For reclamation solutions to be effective, it is necessary to start the collection of salvaged parts early in order to have a well-stocked pool of needed parts (ZVEI, 2002).

7.2.4 Alternate Parts

An alternate part is a part with equivalent or greater performance than the part it replaces (Rogokowski, 2007). Equivalent parts can be provided by another manufacturer or supplier. These parts may be interchangeable with the original specified parts if the functionality is equivalent, the mechanical characteristics and dimensions are the same, and the quality is the same. Alternate parts with higher performance can be used as successors to the original parts, making the older ones obsolete (such as JANTX in place of JAN). In addition, using military temperature range parts in place of commercial lower temperature range parts is defined as using an alternate part (McDermott et al., 1999).

7.2.5 Part Substitution

Part substitution refers to the process of selecting a part as a replacement for an obsolete part. A substitute part may or may not be physically different, but often the substitute part's performance is less capable than the part it replaces. Substitution includes form, fit, and function replacement and, for example, repackaging a modern solid-state component in the same or equivalent package type as the obsolete component (Livingston, 2000).

The 3 F's—form, fit, and function (FFF)—describe the criteria required for a component designed as a replacement part (Jones, 2008):

- The "form" characterizes mechanical attributes such as shape, dimensions, and the ability to withstand or survive the assembly process. In addition, the form defines visual parameters and the possibly haptic characteristics that identify the part.
- The "fit" identifies the specifications needed by the device and describes the suitability to meet the system-level requirements. Parts must have performance characteristics, such as thermal, electrical, and stability characteristics adequate to meet the requirements of the system. In addition, reliability, operational, and environmental parameters have to be achieved.
- The "function" addresses the equality of performance during operation. Function is primarily concerned with replicating input, output, and supply characteristics. However, function also includes the replication of the same bugs, trips, and traps that were present in the original part.

Sometimes there is a fourth F, "finance," which represents the cost difference between the original and the substitute part. For some applications, a chosen part may not be an acceptable substitute if the cost for purchasing or operation is significantly higher than for the original part (Jones, 2008).

One method of mitigating the obsolescence risk is to select parts in the product design phase that have readily available FFF substitutes. Often manufacturers release lists of FFF replacements for parts they plan to discontinue. A cross-reference and comparison of original versus substitute part characteristics must be conducted. In addition, for electronic parts, the approval of the customer (at minimum) is generally required to support the change, since it may require relaxing part specifications or performance parameters. In extreme cases, requalification of the system that the part will be in may be required.

A part characteristic profile should be developed through collection and analysis of engineering, logistics, and technical information during the first step of obsolete part analysis. Additional information should be compiled using the manufacturer's documentation. Identification of all critical parameters and characteristics should be carried out to allow comprehensive evaluation of substitute candidates. The steps in selecting the part substitution process are outlined as follows:

1. Component databases and manufacturer datasheets and websites should be researched to develop a list of potential substitutes for an obsolete part. A list of parts that most closely resemble the obsolete part should be compiled.
2. For each substitute candidate, analysis should be conducted to determine whether the substitute part matches the required functionality specifications and is of reasonable form and fit to be a replacement for the obsolete part. The characteristic data for each substitute candidate part should be recorded and compared with the obsolete part baseline profile. If the part has electrical performance parameters, it should be determined whether these parameters will remain unchanged at the circuit interface where the substitute part is being used. In some cases, electrical performance parameters outside the specified range are important for the correct functionality of the substitute part at the circuit (such as parasitic attributes).
3. When researching potential substitutes, attempts should be made to cross-reference obsolete part manufacturing and functional technology with state-of-the-art product specifications in order to minimize the obsolescence life cycle risk of any substitute selected. The procurement lives of the substitute parts should be evaluated to indicate the life cycle stage and the number of years the part is expected to remain commercially available.
4. A cost-benefit tradeoff analysis should be conducted to determine whether redesigning or design refreshing (see Chapter 9)[4] a product affected by obsolete parts is better than using substitute parts. The cost of support for each substitute candidate and the time frames for part review, approval, and availability should be projected. Cost factors should include:
   - Engineering analysis
   - Source qualification requirements (as necessary)
   - Cost of substitute items (including testing)
   - Associated engineering and logistics documentation changes
   - Requalification (especially for mission and safety critical systems)
5. The approval of the substitute part should be based on:
   - Form-fit-function requirements

[4]Design refresh refers to changes that "have to be done" in order for the system functionality to remain useable. Redesign means "want to be done" system changes, which include new technologies to accommodate system functional growth and new technologies to replace and improve the existing functionality of the system (Herald, 2000).

- Part testing or screening requirements, part manufacturer evaluation
- Module- and end-product–associated integration testing requirements

6. Development and approval of associated engineering changes. Deviation notification should be ensured.

7.2.6 Uprating

Uprating is the process of assessing the ability of a part to meet the functionality and performance requirements of applications where it is necessary to use the part outside of the manufacturer's specification range. Uprating is a special category of part substitution.

The most common form of uprating is thermal uprating. Thermal uprating is used to assess the capability of a part to meet functionality and performance requirements when it is used outside of the manufacturer-specified temperature range, usually at a higher temperature (Condra et al., 1999; Humphrey et al., 2000; Condra and Das, 2001; Das et al., 2001). The technique needs to be applied within the context of an overall parts selection and management plan to ensure the quality and reliability of the resulting product (Jackson et al., 1999; Pecht, 1995). There may be product liability issues arising from this practice, since the equipment manufacturer assumes responsibility for the performance of the part (Biagini et al., 1999; Okey, 1999).

In many cases, part manufacturers discontinue parts rated for wider temperature ranges, while parts with the same functionality but rated for a narrower temperature range remain available. Parts with a narrow rated temperature range are less expensive to produce and qualify and the quantity ordered is higher. In some cases, alternative parts may be available from sources that are rated only for a narrower temperature range. If substitute parts are uprateable, then they can be used to resolve part obsolescence issues.

If a candidate substitute part is available, it is necessary to ascertain the capability of the electronic parts to operate beyond its datasheet specifications, for example, temperature or voltage range.

The first step in uprating is to evaluate the part for uprateability. This is performed through analysis of available temperature versus electrical characterization data. This data may come from the manufacturer, other users, test laboratories that performed characterization of parts, research laboratories, and other sources. If such data exists and is judged to be useable for determining the validity of part performance over an extended temperature range, then the equipment manufacturer can use that data for evaluating uprateability. The equipment manufacturer should continue with the uprating procedure only if the test results indicate that the part may be uprateable.

The next step in uprating is electrical testing for key parameters over the application temperature range to ensure that the parts meet the manufacturers' electrical datasheet limits over the complete application temperature range. The part can be uprated only if a large margin exists between the measured values of electrical parameters and the manufacturers' electrical parameter limits over the entire application temperature range for all of the key parameters. To provide additional confidence that the parts will meet the electrical specifications over the application temperature range, further electrical testing beyond the application temperature range should be performed. The degree of extension of the test temperature range beyond the application temperature range depends on the accuracy level of the predicted thermal environment and the confidence level required for the electrical performance. The part is considered successfully uprated if the results of these tests provide confidence that the part performance will be stable over the application temperature range. If it is concluded that the part cannot be uprated, then other alternatives need to be explored. These alternatives include thermal management, performance derating, and application derating.

Once an electronic part is uprated and is inserted in a system, then the supplier enters a cycle of system- and component-level testing to ensure continued performance. Another input for continued monitoring is a process change notice from the manufacturer. If the effects of a process change notice are believed to impact the component's performance with a change in temperature, then performance testing should be redone. Otherwise, system-level and sample testing will be continued.

Often, no physical design modification of the board is necessary to introduce an uprated part to a system. If the part is from the same manufacturer and part family as the obsolete part, then the additional part manufacturer or part family qualification will not be necessary.

The process of uprating can be expensive if it involves significant analysis and testing. There is also the potential for legal issues if parts are used beyond their ratings because the part manufacturers are not responsible for any part failures and all warranties are void.

A part that is uprateable now may not remain so in the future,[5] and hence uprating may not be a long-term solution. It is necessary to develop and maintain a continued monitoring process for the use of uprated parts. Such a process will add cost to the use of uprated parts. Some customers may not accept the use of uprated parts in their systems due to legal or contractual reasons. Depending on the type of uprating methodology used, the electrical design of the system may have to be altered, and that may limit the use of uprated parts in some situations.

As an example, Honeywell Engines and Systems faced an obsolescence problem with its Full Authority Digital Engine Controller (FADEC) applications when Motorola discontinued its military product line. The military temperature range (–55°C to 125°C) version of the Motorola MC68332 microcontroller was being used earlier in this application, and Honeywell needed a replacement for the discontinued part. An equivalent Motorola part with the same functionality, MC68332ACFC16, rated for a –40°C to 85°C temperature range, was selected and assessed to replace the discontinued part.

[5]The manufacturer of the part may change process or other parameters (for which a PCN may not be issued) that affect the uprateability of the part.

As uprating is a special category of part substitution, a cost-benefit tradeoff analysis should be conducted, as described in Section 7.2.5, to determine whether redesigning a product is better than using an uprated part.

7.2.7 Aftermarket Sources

Aftermarket refers to the period after the original manufacturer has phased a part out of production. Aftermarket sources cater to the continued demand for discontinued parts. Three types of aftermarket sources exist: approved aftermarket sources that provide finished parts, approved aftermarket sources that remanufacture parts, and unapproved aftermarket sources for finished parts (brokers). Approved aftermarket sources have formal agreements and approvals from original manufacturers to resell their parts in the aftermarket.

If part specifications, test results, and related technical data are complete and available, aftermarket manufacturers may support continued production of obsolete items. Some aftermarket sources for electronic chips also buy and store wafers and can package die to suit the customer.[6] Part manufacturers often establish transfer agreements with aftermarket sources for discontinued product line support.

Industry and trade groups such as the SIA[7] endorse using parts from aftermarket sources. SIA states, "Increasingly, major semiconductor suppliers are incorporating qualified and authorized aftermarket sources into their customer support programs." In many cases, these qualified aftermarket sources can be a system manufacturer's first and most cost-effective option for resolving its obsolescence problem. The emergence of dedicated aftermarket sources having formal relationships with the original IC manufacturers represents a viable means to support continuing requirements beyond last-time buy opportunities.

Formal relationships between major part manufacturers and aftermarket sources have flourished in recent years. In many cases, product discontinuance notices and press releases include the names of authorized aftermarket sources. This practice can reassure suppliers that the major manufacturers have confidence in these sources. For example, Texas Instruments has designated the aftermarket manufacturer Rochester Electronics as its authorized distributor of all obsolete parts. This authorization includes the transfer of responsibility for product support (Texas Instruments, 1999f).

Just because aftermarket sources are approved by the original IC manufacturer does not necessarily guarantee that aftermarket parts are of the same reliability and quality as the original parts. The confidence of equipment suppliers in aftermarket sources depends on whether these market sources maintain a continuous supply of parts after the original manufacturers have discontinued them.

[6]Approved aftermarket sources can usually only package die in exactly the same package as the original part.

[7]The Service Industry Association (SIA) is a nonprofit association of enterprises involved in the support of high-technology products in the marketplace.

When considering aftermarket alternatives, the obsolescence management team should evaluate the aftermarket source's storage facilities and practices. If the aftermarket source is performing any manufacturing activities, the manufacturer's production capabilities and test programs should be evaluated to ensure the ability to meet original item specification requirements. Acceptance tests and qualification requirements for the obsolete part should be identified. The adequacy and accuracy of technical information from the part datasheet should be validated.

Some aftermarket sources distribute catalogs that identify the products and technologies they support and provide detailed cross-references to original part numbers. The original part manufacturers may be contacted to assist in locating suitable (approved) aftermarket sources. Often, original part manufacturers distribute lists of cross-referenced parts together with suitable aftermarket sources or even other manufacturers/competitors who make similar parts. Once identified, supplier capabilities should be evaluated to determine if they meet original item production and qualification requirements, and production plans for the expected duration of obsolete part procurements should be assessed. To accomplish this objective, part technical data and part test and qualification requirements need to be provided to aftermarket sources to assess the ability to manufacture form, fit, and function (FFF) replacements for the part in question. If aftermarket suppliers are not qualified to the level of the original part specifications, an engineering analysis may be appropriate to determine the potential for waiving selected requirements to allow approval of the aftermarket part. Alternatively, the feasibility and cost of qualifying supplier facilities to the desired level should be evaluated.

To evaluate an aftermarket manufacturer, bids are requested from aftermarket sources for designated quantities of obsolete items. Part production and delivery schedules should also be determined. The total cost of an aftermarket resolution should include:

- Technical data procurement/development
- Qualification of obsolete parts
- Manufacturer (aftermarket) assessment
- Cost of the obsolete part (including testing)
- Documentation changes if engineering change proposals (ECPs) are required
- Material handling and storage if volume procurements are foreseen

The time frame for implementation of this alternative should be determined from in-house procurement lead time through item qualification and material delivery.

The companies that provide aftermarket services vary greatly in size, capabilities, and experience. It is expected that these variations are reflected in the cost, reliability, and, most importantly, quality of their products. This lack of consistency is one of the reasons many OEMs are reluctant to rely on these companies for purchasing obsolete parts. As a relatively new and seemingly lucrative business opportunity, the aftermarket has attracted many organizations, and while many of them are competent and capable, others are not. Some OEMs have had less than satisfying encounters with aftermarket companies and have found that aftermarket parts do not always match the performance of original parts (Pecht et al., 1997). The availability and lead times to receive aftermarket parts vary significantly. Some aftermarket companies have national (and international) distribution networks, while others operate from a single location.

Capabilities for quality monitoring and testing vary widely among aftermarket companies. Another difficulty is that, unlike major manufacturers, the availability of detailed quality data on these firms is limited. While some issues are more relevant for choosing aftermarket sources than for choosing regular suppliers, the manufacturer part family and distributor assessment criteria should be used to assess aftermarket sources. When the aftermarket source does the design, manufacturing, or packaging of discontinued parts, the manufacturer's part family assessment (Syrus et al., 2001a; 2001b) criteria should be used. When the aftermarket source performs inventory management for original manufacturers, distributor assessment criteria should be used. Due to small production volumes, it might be hard to determine the composition of aftermarket part families. Just as when buying from original manufacturers, equipment suppliers must be knowledgeable about the particular strengths and weaknesses of aftermarket sources.

The number of aftermarket companies is increasing as more organizations identify the aftermarket as a business opportunity. Aftermarket companies in the electronics industry offer four main manufacturing services for discontinued parts, each requiring different levels of technological expertise:

1. Trading:
   - Clearing house for excess inventory: Electronic data transfer has made clearinghouse activities a viable process. Clearinghouse type aftermarket companies act as intermediaries between organizations holding excess inventory and those with a need for discontinued parts. With many government and military agencies offering similar services free to their OEMs, it is difficult for private firms to compete in this market sector.
   - Sourcing agent to locate inventory: Some service companies act as buying agents for organizations facing the obsolescence problem. Contract assemblers are well qualified to conduct a search for obsolete parts, as they have developed a wide range of contacts and expertise.
2. Distributing:
   - Inventory of discontinued parts: This practice is different from the trading business, as it involves the risk of holding inventory. It also requires proper storage and handling facilities for managing high-reliability parts. Firms' levels of involvement in this process vary substantially. Purchasing inventory can be a collaborative process

between original manufacturers and aftermarket sources, or it can be a unilateral effort on the part of aftermarket sources only.

Some companies monitor technology trends and buying patterns along with road maps of individual manufacturers and industry associations. Based on that data, they acquire an inventory of parts likely to become obsolete, and they stock the parts for future sale. Other companies team with original manufacturers and become authorized distributors of the manufacturers' discontinued parts. This practice is becoming more common, and many companies now include the names of authorized distributors in their product discontinuance notices. Some of these agreements are for individual products; the trend, however, is toward maintaining ongoing, long-term contracts, which give the suppliers some assurance of stability.

Some traditional distributors and wholesalers of electronic parts are also venturing into the business of distributing obsolete parts. Because of their expertise in customer interface, they often take on the complete business aspect of the end-of-life (EOL) process. This work involves customer notification, inventory handling, order processing, and lifetime-buy fulfillment (Elantec, 1997).

- Inventory die and wafers: This business is similar to purchasing and reselling inventory; however, managing die and wafers requires special handling capacities. Most of the companies in this segment were originally die and wafer processors that have expanded their businesses to cover purchasing and stocking in response to the obsolescence problem. Some companies stock die as part of their agreements with manufacturers to handle all discontinued inventories. Other companies concentrate on this area.

3. Assembly and Testing:
   - Repackage bare die to meet extended specifications: Some repackaging companies supply products from their stock of die and wafers bought from original manufacturers or authorized distributors, while others take individual orders for die packaging. Some companies offer subsequent testing and qualification services for parts slated for use in harsher environments.
   - Test and certify parts to meet extended specifications[8]: Companies established in testing and qualification have ventured into the market for extending the temperature-range (uprating) certification of electronic parts. However, there are relatively few of these companies. One reason is that many part manufacturers strongly discourage this practice. Another is that the practice of using parts beyond manufacturers' specifications may entail legal liabilities in the event of system failure (Kroeger, 1997).

[8]Some suppliers perform test procedures themselves to mitigate the risk of using parts beyond their manufacturer-designated limits through uprating and performance derating. Any strategy to handle obsolescence must consider developing and using these techniques.

4. Design and Manufacturing Businesses:
   - Operate discontinued product lines: The practice of buying and operating entire discontinued product lines that still have potential for mid- to long-term sales is gaining momentum. Often, product lines that are only marginally profitable to the original manufacturer can still be a good business opportunity for smaller and niche-market operators. Such manufacturers purchase and gain rights to fabrication lines, designs, mask sets, testing software, and equipment from the original manufacturer and continue production. This trend is present for both military and non-military grade older technology components.

     Numerous companies are involved in aftermarket manufacturing, and even more companies are expressing interest in this segment. With the removal of "onshore" clauses from the U.S. military qualification guides, some Asian manufacturers are also exploring this business opportunity.
   - Design and manufacture equivalent parts[9]: Some companies design and provide alternative parts for discontinued technologies. Usually these companies concentrate only on their established areas of expertise. Without gaining intellectual rights from the original manufacturers, this method is realistically applicable only to simple digital parts.

Equipment suppliers have raised a number of concerns about purchases from aftermarket companies. The major concerns are cost, quality, and counterfeit risk. For them, testing is not only essential to prove functionality, it is also necessary to protect against counterfeit parts.

Counterfeit refers to all different forms of imitations, forgeries, copies, and misrepresentations (Battersbury, 2008). There are many components for which there are known counterfeits that have the potential to create widespread and serious problems. Counterfeiting of electronic components started when the majority of electronic products were expensive to develop and qualify for military applications, with unreliable sources failing to conduct all mandatory testing (Rochester, 2007).

Counterfeiting operations can sell their parts when there is a manufacturing shortfall/product shortage, an obsolete/discontinued device, or any other demand for high-value products.

Today, sales of counterfeit components and replicas are increasing worldwide (Rogokowski, 2008):

- In 2005, up to US$200 billion of international trade was in counterfeit or pirated products (Organisation for Economic Cooperation and Development [OECD], "The Economic Impact of Counterfeiting and Piracy").

[9]Specialized reverse-engineering, design retargeting, and emulation companies are not included in this category and are discussed separately in later sections.

- In 2006, up to 10 percent of technology products sold worldwide were counterfeit, amounting to US$100 billion of sales revenues (Alliance for Grey Market and Counterfeit Abatement [AGMA]).
- In 2007, it was estimated that counterfeiting activity drains about US$250 billion out of the U.S. economy each year. In 2007, 9 percent of the total value claimed by counterfeiters was in consumer electronics (U.S. Patent and Trademark Office).
- It is estimated that counterfeiting accounts for between 5 and 7 percent of the value of world trade (International Chamber of Commerce).

There are many ways that counterfeit devices can originate, such as the manufacturing of a completely fake device, inadequate control during disposal of scrap parts and rejects, reclamation of used components, rebranding to sell at higher cost, and false claims of conformance with regulations, such as the restriction on hazardous substances (RoHS) (Rochester, 2007).

Generally, there are significant quality concerns about counterfeit parts, because they can introduce potential weaknesses, incomplete or inadequate functionality, performance limitations, and may fail earlier than expected. In addition, the cost impact on the original component manufacturers (OCM) is potentially high. OCMs may realize lower earnings because of the low-cost counterfeits available on the market. This will affect the ability of the OCMs to generate new products and produce returns for their shareholders (Rochester, 2007).

As described previously, it is necessary to implement effective anti-counterfeiting tactics. The following list presents some tactics that are now being adapted by leading electronics companies (Rochester, 2007; KPMG, 2005; Battersbury, 2003):

- Visual testing of the parts, forms, dimensions, and labeling
- Checking certificates of conformity, specifications, and delivery notes of the parts
- Designing and employing copy-resistant and anti-counterfeit technologies
- Contacting special organizations regarding specific knowledge, machinery, and equipment for quality inspection of suspect parts
- Coordinating with U.S. and international organizations and trade alliances
- Partnering with international law enforcement and customs agencies
- Contacting the original manufacturer of the part for assistance
- Developing and deploying anti-counterfeit education and training programs for internal and external stakeholders
- Ongoing monitoring of distribution streams of counterfeit products

Many equipment suppliers tend to trust the product quality of high-volume manufacturers over that of aftermarket sources. High-volume manufacturers generally have superior processes to control yield, quality, and reproducibility compared to parts manufactured by relatively small aftermarket sources.

In some instances, the aftermarket vendor may take on the appearance of a brokerage house. But the differing management styles and business techniques employed in this niche market are truly astonishing. Fierce competition in industry forces the aftermarket firm to choose a product wisely, generally selecting a product with known return on investment (Easter, 1994).

Aftermarket companies claim that their parts are priced at about the same levels as original parts, but many buyers complain that prices of parts from aftermarket sources are much higher than originals and carry minimum order quantities. This is not surprising, since demand and production levels are both lower than when the part was originally produced. Some price differentials reflect the fact that original manufacturers sold the products at or below cost when they were in the market. The prices of parts also tend to escalate with time. The price differential depends on the method by which the parts are supplied.

Aftermarket companies that depend on the inventory of original manufacturers will inevitably run out of stock at some point, and another part shortage will ensue. Although aftermarket companies can remain viable for smaller production volumes than original manufacturers can, they need some minimum volume for their product lines to be viable. There is no guarantee that in the future there will be enough of a market for them to be able to continue to produce the parts economically. When aftermarket product lines are closed, new obsolescence problems occur.

Repackaging companies that buy commercial and industrial die from original manufacturers may not be able to supply parts with form-fit-function matches if the original manufacturers make major design changes. It is expected that every part will undergo a fabrication facility move, wafer diameter change, electrical design change, die shrink, or combination of these within three years (Kroeger, 1997). Without proper analysis and testing of the altered capabilities of the changed die, only special packaging may deliver extended environmental functionality. More stability occurs when aftermarket sources continue to manufacture the products.

Moreover, aftermarket and original manufacturers enter into exclusive agreements, a practice that can allow critical parts to become single-sourced and be at greater risk of becoming unavailable or higher priced in the future.

Even though the aftermarket has become a parts source in recent years, it has simultaneously hindered the insertion of newer technologies (mostly in the military and avionics marketplace). Policymakers and regulators should reevaluate their parts procurement strategy so that those products do not remain stuck at the trailing edge of technology.

7.2.8 Emulation

Emulation is primarily applicable to electronic parts. It refers to the process of creating unavailable electronic components from their slash sheets, datasheets, test vectors, and other information. Emulated products are sometimes categorized as substitute or alternative parts (they are not aftermarket parts).

The foundry best known for developing and supplying obsolete microcircuits by emulation is Generalized Emulation of Microcircuits (GEM). The GEM BiCMOS process (Crane, 2002), the GEM Design System and Library, and the specification procedures, packaging, interfaces, and testing procedures were developed under the GEM R&D program. Started as a United States tri-service manufacturing science and technology R&D initiative sponsored by the Office of the Secretary of Defense, GEM is currently managed by the Defense Logistics Agency (DLA).

The GEM process recreates unavailable electronic components and allows continuous manufacturing of microcircuits in a quality-controlled environment. The customization process occurs at the last stage of creating interconnects on the emulated wafers to create the die. Figure 7-3 schematically shows the process flow of IC emulation.

Emulation is currently only possible if the microcircuit in question meets specific physical or functional characteristics regarding the following:

- Technology type (RTL, DTL, TTL, NMOS, PMOS, and so on)
- Operating voltage
- Temperature range
- Pin and gate count

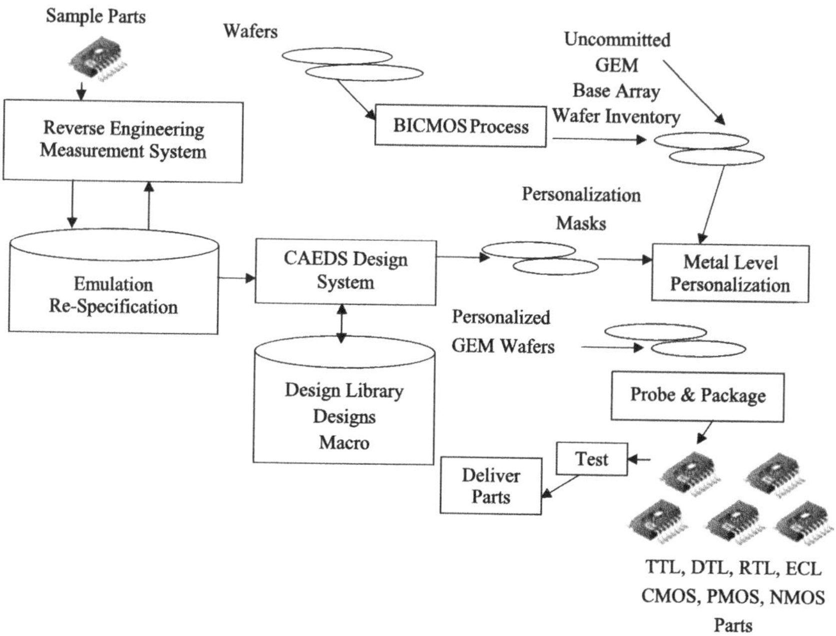

FIGURE 7-3 The GEM emulation process (Crane, 2002).

Contact with prospective emulation activities will generally quickly determine whether the part is an emulation candidate.

The availability and adequacy of part technical data and specifications, including part testing and qualification requirements, to support the emulation process should be assessed. Accurate and complete part specifications are essential to ensure that the emulated product matches all of the characteristics of the original part. This specification could be developed from existing government and commercial documentation, as well as from measurements made of existing parts to clarify questionable data. Coordination with original part manufacturers to obtain final acceptance test procedures or other specification requirements may be necessary. In some cases, failures and bugs that are in the original part also need to be emulated to ensure the equivalent functionality of the emulated part of the circuit.

Candidate emulation firms should be identified and the technical data necessary to conduct emulation feasibility analysis (as required) should be provided. The acceptability of the proposed emulation approach and associated material production and delivery schedules should be validated. As in the aftermarket alternative, emulation firms may not maintain item qualification capabilities to the level of the original part specifications. The potential for waiving specifications or funding the qualification of the potential source may need to be evaluated.

Bids from selected emulation activities for designated quantities of obsolete parts may be requested, ensuring consideration of varying order quantities.

The total cost of this alternative should be assessed, including:

- Technical data/specification development
- Qualification of obsolete parts
- Manufacturer (aftermarket, emulation agency) assessment
- Part test/acceptance costs (including full-system integration testing as required)
- Engineering/logistics data revisions
- Material handling/storage (if volume procurements are anticipated)

The time required for completion of the emulation process, including in-house lead times and engineering change package analysis and approval, should be estimated and recorded.

There are certain advantages to using emulated parts. They generally provide form-fit-function[10] replacements of the original parts. Emulated microcircuits often also have a complete tracking history and meet all the specifications of the original parts, including physical (package type, pin out),

[10]Matches all datasheet and selected item parameters, but complete manufacturer test vectors are not generally guaranteed.

electrical (functional DC, dynamic, power), and environmental (electrostatic discharge, temperature) parameters. Usually, emulated parts can be supplied in small lots because a common wafer is used for all products. Manufacturing can be done on demand, reducing the need for high inventory and allowing for mistakes in estimating the quantities of required parts. It is also possible to tailor parameters for special applications, so failures can be reduced by using specially tailored emulated products with adjusted parameters.

The prototype development time for new emulated parts can vary from 60 to 90 days to get an emulated part that is already in the catalog and up to 12 to 18 month for new parts that are not in the catalog. The program also offers stability, as it has only one product line and is not likely to close or discontinue a specific product. Nonrecurring expenses (NRE) depend on whether the same part or similar parts have already been emulated.

Not all obsolescence problems are solvable through emulation technology. The microcircuit technology range is limited. Limitations are given, for example, by technology generation, pin count, gate count, and propagation delays. The price per unit of the emulated microcircuit is extremely sensitive to order quantities and the complexity of the part.

Because it is not possible to emulate all parts, and emulated part prices can be prohibitive, many suppliers have looked for other solutions to obsolescence that can cover a larger number of parts. This explains why the total number of emulated parts available and under development is still low, even though its capability has been demonstrated in a large number of systems. Some skeptics of the emulation program advise equipment suppliers to rely on emulation only as a "safety net."

Currently, the focus of emulation programs is almost exclusively on military parts. Unless emulation is able to respond to the needs of the consumer marketplace and supply advanced technology parts, its long-term viability is not ensured.

For example, emulated parts have been used successfully in programs run by the U.S. Air Force (F-15, F-16, JointStars), Army (SINCGARS, MLRS), NASA (Shuttle Orbiter), Navy (A-6, Aegis), and many others. The U.S. military services claim major cost and time savings from using emulation technology in these programs.

Emulation is also a reactive strategy to rebuild obsolete software. In some cases, the original vendor may allow the customer to purchase the source code for the product and thus enable the customer to emulate the software. Furthermore, a third party can be contracted to emulate or continue support of an obsolete software application (Sandborn, 2007).

7.2.9 Redesign

Redesign involves designing out obsolete parts via engineering changes at various system indenture levels with a view to enhancing system performance, resolving part obsolescence, and improving reliability and maintainability.

Redesigns at the component level may involve significant risk if the item in question has different applications, and extensive system integration testing and qualification may be required. Moreover, depending on the scope and level of the redesign effort, substantial nonrecurring engineering (NRE), requalification, and life cycle logistics costs may accrue. Given the relative magnitude of engineering and logistics cost factors, redesigns may be most appropriate when a large fraction of current or potential obsolete parts are resident within a particular system or end-item. The cost of redesigning to replace an obsolete component depends on diverse factors, especially if the product affected by the obsolescence of a part has other vintage technology parts that are obsolete or are soon to be obsolete. In this case, redesign might be more cost effective than other options.

A redesign should be explored as a reactive solution only after evaluation of the other alternative solutions mentioned in this chapter, such as selection of an interchangeable replacement component or an alternate source for a component.

Usually, redesign is the last option, as it is perceived to be expensive, but there can be situations when it is viable, both economically and technically, depending on two major factors: the usage patterns of the components in question and the time available for redesign. However, redesign may be a viable alternative for strategic management of obsolescence because it can resolve multiple current obsolescence issues and potentially avoid future obsolescence issues (see Chapter 9).[11] Component usage patterns determine how many redesigns must be done and who is responsible for them. Usage patterns also determine the requalification process, if any, that the new components will have to undergo. The time available for redesign depends on when the components are projected to become unavailable. In considering redesign, it should also be seen as a marketing opportunity for functional additions, cost reductions, and reliability improvements. This may be the only viable option once the last-time buy parts have been consumed.

Ultimately, all long-life products will require redesign, since the other options are all temporary solutions intended to overcome an immediate problem. It should always be considered that long-life systems will be in a constant state of redesign throughout their lives.

Any redesign process undertaken to solve an obsolescence problem must address future availability concerns to ensure that the new design has a longer, more useful, and more maintainable life. Another consideration during the redesign process is ensuring that the best currently available and easily upgradeable design and manufacturing technology is used in the new designs. The goal of redesign should reach beyond just solving the immediate problem of the nonavailability of parts (see strategic solution discussion in Chapter 9).

[11]Although often used interchangeably, in general a distinction is made between design refresh and redesign. Design refresh refers to changes that "have to be done" in order for the system functionality to remain useable. Redesign means "want to be done" system changes, which include new technologies to accommodate system functional growth and new technologies to replace and improve the existing functionality of the system (Herald, 2000).

The steps in the redesign process are as follows:

1. The status of current and planned redesign initiatives for cognizant systems/products should be assessed and associated technical and cost proposals submitted.
2. Redesign engineering and cost proposals should be reviewed and approved; feasibility studies, design and cost tradeoff analyses, and any other information necessary to validate the schedule and the technical and financial aspects of the redesign approach should be conducted.
3. Regarding the first objective, analysis of both existing and potential obsolescence problems on a total system, product, or end-item basis may be necessary to determine the most efficient indenture level for redesign efforts. Design changes should be incorporated on a sufficiently broad scale to minimize continued part obsolescence impacts (modular designs). At the same time, parts used in the final engineering change package should also be screened to determine the obsolescence risk.
4. Cost projections for this alternative should be developed, including:
   - Up-front system engineering, testing, procurement, and installation, as well as technical data development and modification.
   - Life cycle logistics supports the costs (both manpower and material) associated with the redesigned item. A number of available logistics cost-estimating models, which provide product life cycle cost factors for product engineering changes and alterations, are available to assist in this analysis.
5. Execution time frames required for completion of the redesign process should be determined.

There are some advantages to using redesign as a solution for component obsolescence. A redesigned system using newer technology and design tools may be less expensive to manufacture and maintain and may achieve better tolerance on electrical parameters than the design it is replacing. Redesign also provides the possibility for technology insertion and other design improvements, such as cost reductions, reliability improvements, or additional functions.

The chief limitation of redesigning a system is that the original design team has often been disbanded so that the post-design team has to rely on drawings and documentation. This point emphasizes the importance of documentation and strict drawing configuration control in enabling the sustainment of a system over a long period. For a redesign, requalification of the system (including system software) may be necessary, and compatibility with the rest of the system may be an issue.

In summary, redesign can usually be classified as either minor or major. Minor redesign refers to the re-layout of a system or a circuit card assembly, whereas a major redesign refers to the total replacement of certain parts of a system or of a circuit card assembly (McDermott et al., 1999).

7.2.10 Reverse-Engineering

Reverse-engineering is the process of developing exact replicas of candidate items through the review of available technical data, physical disassembly, and analysis of item components. Reverse-engineering may be appropriate when sufficient technical data or data rights to support re-procurement are unavailable. The goal is to cultivate qualified alternative sources. The reverse-engineering process is generally expensive and may be used for cost or technical comparison with redesign or other solutions.

The steps in selecting the reverse-engineering process are outlined as follows:

1. Part technical and procurement data availability should be ensured, together with the availability of sample parts for analysis and disassembly during the reverse-engineering process.
2. Industrial or government organizations that may be able to reverse-engineer the part should be identified. Historical contract information and technical data should be provided to the prospective reverse-engineering organization. Feasibility analyses, technical/cost proposals, completion schedules, and plans for incorporation of product improvement or redesign initiatives as part of the reverse-engineering process should be conducted.
3. Costs for the reverse-engineering alternative should be estimated, including:
   - Reverse-engineering feasibility analyses and studies
   - Source set-up and tooling
   - Prototype production
   - Re-procurement data package development
   - Part-level qualification
   - Review and approval
   - Material costs
   - Additional product life cycle logistics manpower/material costs resulting from redesign or product improvement initiatives
4. Reverse-engineering execution time frames should be determined and documented.

For example, an emulated part or IC is not an exact replica of the obsolete part, whereas reverse-engineering is an attempt to create an exact replica through review of available technical data and through measurement, testing, and physical disassembly of the part. This is usually very expensive and can require large amounts of time. In addition to the obvious advantages of being able to manufacture a form-fit-function replica of an original electronic part, newer packaging technologies may be used to meet today's increasing demand for smaller package sizes.

Reverse-engineering, like redesign, is an expensive process that requires specialized technical knowledge and infrastructure. Furthermore, issues arise about the intellectual property rights of the original manufacturers. Only if the customer (often the government) or supplier has the right to the technology in question can these options be exercised.

In one case, the U.S. Army reported a success story whereby reverse-engineering was employed to resolve an obsolescence issue. When the original manufacturer of a hybrid IC used in an Army communications system no longer supported the part, the system manager, to avoid the prospect of costly redesign, called upon the DMEA engineers to reverse-engineer the obsolete IC. From the reverse-engineered IC, a prototype was created that was tested in the communications product. After sufficient proof that the reverse-engineered IC worked exactly like the original obsolete version, the DMEA finalized the design and developed a physical layout that could be cost-effectively manufactured in small volumes. The final parts were tested, qualified, and delivered to the army for continued support of the communications system. This approach saved the army millions of dollars and years of wasted time (Pecht et al., 2002).

7.2.11 Lifetime Buys/Bridge Buys

In a lifetime buy (also called life-of-type or LOT buy), the equipment supplier buys sufficient parts to meet the system's lifetime needs from the original part manufacturer before the anticipated termination of production. This is one of the simplest solutions because it does not require reengineering, requalification, or redesign, and it is one of the first options considered for mitigating obsolescence. Part manufacturers also promote this option when they plan to exit the market for a specific product or line.

A bridge buy is a buy made to meet future demands until the next redesign or modification of the system that results in the replacement of the obsolete part. A typical time line for the usage of stocked parts before and after a bridge buy is shown in Figure 7-4. As a bridge buy is a subset of a lifetime buy, it has the same advantages and disadvantages explained in this section.

The equipment supplier should determine the last dates for order processing, minimum order quantity or value, and the latest delivery dates for the part from the manufacturer's product discontinuance notice (PDN).

Lifetime buy costs should be calculated as a baseline for comparison with other options, and the lifetime buy execution time frame should be determined and recorded:

- Total material costs should be calculated by requesting a quote from the manufacturing sources based on the total quantity required. Item testing requirements should also be considered as part of the total cost.
- Overhead costs should be calculated. They should include storage and handling, inspection, testing, repackaging, and shipping. When special

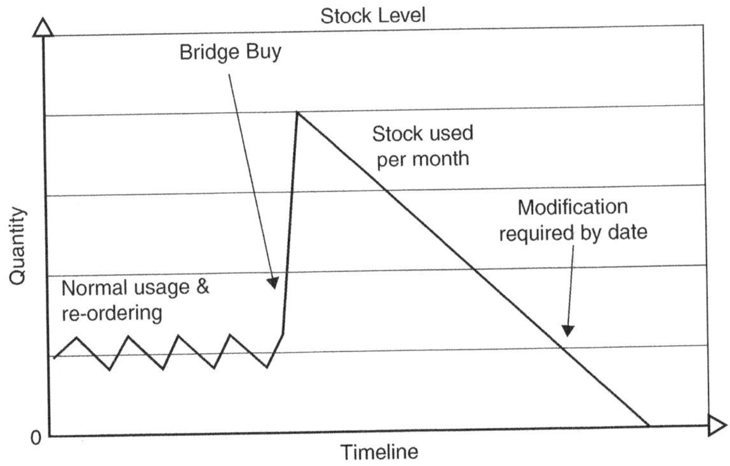

FIGURE 7-4 Bridge buy stock graph (adapted from Hill, 2007).

storage conditions are required, such as dry nitrogen storage, they should also be factored in.

- Cost of money. Lifetime and bridge buys require the purchase of parts earlier than they would otherwise be purchased. If the effective discount rate on money is non-zero (it almost always is, even for governmental organizations), then the cost of money associated with lifetime and bridge buys must be included.
- Cost of buffers. Lifetime and bridge buys often purchase more parts than the expected demand (buffers). The cost of the buffer part purchases must be included.
- Total lifetime costs are calculated by totaling the material, overhead, buffer, and money costs.

For electronic parts, especially microcircuits, the presence of die banking and repackaging houses results in some suppliers reducing the risk and financial commitment of lifetime buys by purchasing die instead of complete parts. The purchase price of a die is less than the cost of a finished part, allowing for the purchase of buffer stocks. Controlled storage of a die can be done at some assembly houses. Even if the original die manufacturer no longer supports a military product line, as long as the die is available from a source that is already qualified for use by the customer or acquisition agencies, suppliers have an economical solution to the obsolescence problem. Assembly and packaging houses can provide final processing services. Die banking also provides the option of using advanced package types and technologies. Storage and repackaging facilities must be carefully chosen to avoid damage to the stock of die.

The option of making lifetime buys of components slated for discontinuance is tempting, because it allows for the use of the same part without design modifications. No new sources are needed, nor is requalification required. Another benefit of lifetime buys is that parts bought directly from manufacturers also come with the original manufacturers' warranties. This may not always be beneficial for customers with long-term storage requirements, as manufacturers' warranties will run out before the part is needed.

A lifetime buy is also beneficial for manufacturers. They receive large and predictable orders that help them to fully utilize the capacities of product lines before the lines are discontinued. For example, the Semiconductor Industries Association Government Procurement Committee (SIA/GPC) has recommended that government acquisition agencies make a contractual provision that allows special payments for bulk purchases of possible obsolete parts (SIA, 1996). The SIA/GPC also proposes that contracting reforms allow for long-term storage of critical components for the entire life of a program and for subsequent spares for repair purposes. SIA moves indicate that the manufacturing industry prefers consumers to accept the option of lifetime buys.

Regarding the maintenance of existing systems, making a lifetime buy of any significant quantity is not only an engineering decision but also a complex financial decision that requires dependable financial forecasts at both the organizational and macroeconomic levels. Some of the cost factors that influence a lifetime buy are shown in Figure 7-5.

The discount rate on money and the anticipated holding time of part inventory play important roles in justifying lifetime purchases. Accurately forecasting the required quantity for a lifetime buy purchase is essential. If the affected program has a "sunset" date and the number of units to be produced is known with certainty, then the number of parts required for manufacturing the units may be predicted. For an active system, when numbers yet to be manufactured are uncertain, it is harder to determine the number of parts required for the completion of the project and to take into account a part's environmental and operating requirements, particularly when the part in question is being used under varying conditions in different products. The replacement part requirements for each system will be different. Common reliability prediction methods used to estimate the number of spare parts needed are often inadequate (Pecht et al., 1997; Cushing, et al., 1993).

There are several models and tools for lifetime buy forecasting. Examples include the CALCE Stochastic Individual Buy Model and the Life of Type Evaluation (LOTE) tool. These models were created to optimize lifetime buy quantities and minimize product life cycle cost. To generate an optimized buy quantity within a specific confidence level, these tools also need at minimum the input of the length of time that companies are buying for (in time periods), the demand forecast in each time period (this can be correlated period to period), the length of time needed to design out the part or identify another solution (if necessary), and the desired confidence level (Feng et al., 2007; Sandborn et al., 2010).

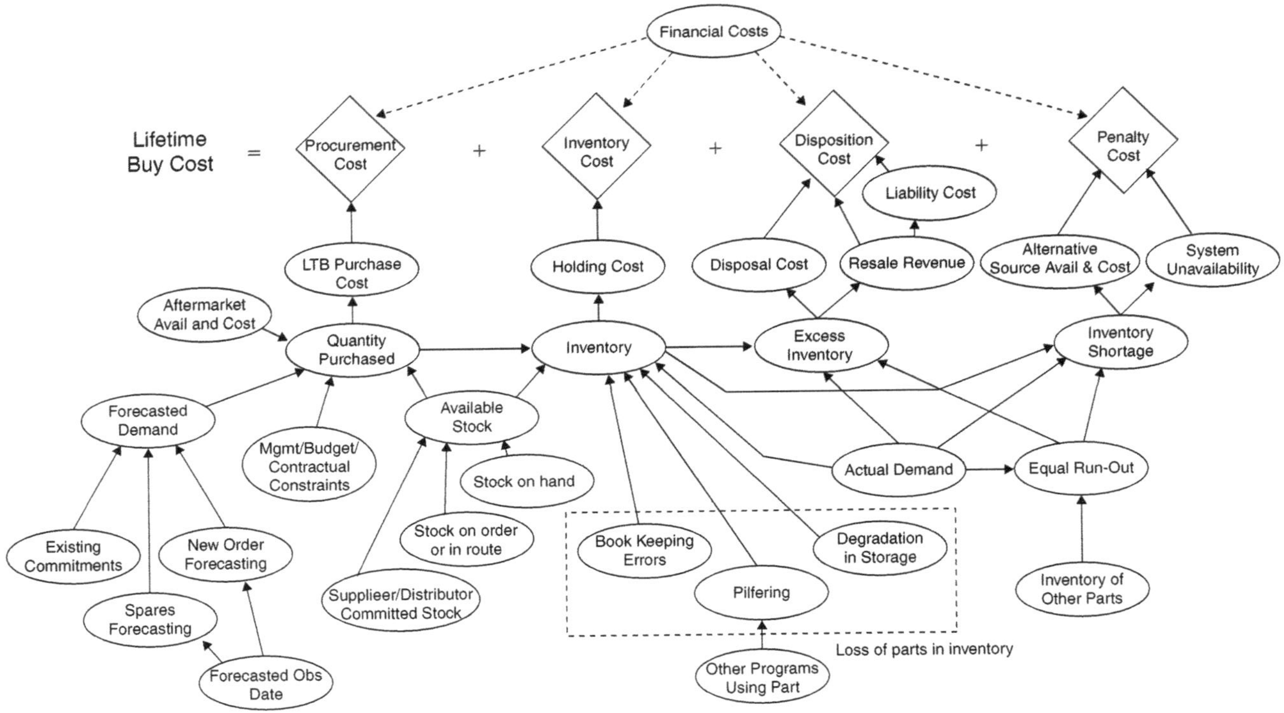

FIGURE 7-5 Cost factors affecting part lifetime buys (Feng et al., 2007).

TABLE 7-2 U.S. Code: Title 10, Chapter 131, Section 2213, Limitation on Acquisition of Excess Supplies (Pecht et al., 2002)

Sec. 2213. Limitation on acquisition of excess supplies
(a) Two-Year Supply—The Secretary of Defense may not incur any obligation against a stock fund of the Department of Defense for the acquisition of any item of supply if that acquisition is likely to result in an on-hand inventory (excluding war reserves) of that item of supply in excess of two years of operating stocks.
(b) Exceptions—The head of a procuring activity may authorize the acquisition of an item of supply in excess of the limitation contained in subsection (a) if that activity head determines in writing:
(1) that the acquisition is necessary to achieve an economical order quantity and will not result in an on-hand inventory (excluding war reserves) in excess of three years of operating stocks and that the need for the item is unlikely to decline during the period for which the acquisition is made; or
(2) that the acquisition is necessary for purposes of maintaining the industrial base or for other reasons of national security.

Even when predicting part requirements can be done with acceptable accuracy and when the cost of maintaining a lifetime inventory is not a problem, a bulk purchase may not be the best engineering decision. To sustain a long-term program with a lifetime purchase, the parts must have proven dormant storage integrity, as there is always the possibility that manufacturers' warranties will run out before the part is needed. If there are failures or if the parts degrade during storage, a poor forecast of the number of parts can mean that the parts will be in short supply once the process lines are closed. The engineering justification for bulk purchase of the part in the first place is thus negated.

Lifetime buys also run counter to "lean" manufacturing practices, such as just-in-time production, which reduce inventory overheads and inventory maintenance costs. The Defense Supply Center spends considerable sums of money in lifetime buys every year and consequently holds many millions of dollars worth of components that will not be used for a long time or that may never be used.

The U.S. Department of Defense (DoD) has to operate in accordance with the Defense Federal Acquisition Regulations (DFARs). In particular, there is a general prohibition on the purchase of quantities of parts beyond immediate need (U.S. DoD, n.d.). The text of this regulation is reproduced in Table 7-2. This requirement places severe limits on the ability of government to use last-time buys as an obsolescence solution by setting a limit on the purchase of spare parts. Even then, the military's extended buys normally cover only requirements for spares and maintenance. Although suppliers can procure parts for new production, lifetime buys for new products must be funded from other sources. Manufacturers are also under no obligation to provide the option of a lifetime buy, and they often do not, due to limited availability.

7.3 SELECTING THE PROPER REACTIVE OBSOLESCENCE MANAGEMENT STRATEGY

The major factors for consideration in choosing a strategy for solving an obsolescence problem are discussed in the following subsections.

7.3.1 Part Discontinuance Status

The issuance time line of a product discontinuance notice is illustrated in Figure 7-2. Prior to the issuance of the PDN, the equipment supplier may be unaware of the impending event unless some forecasting tool is employed. Once the PDN has been issued, the estimation process for the various options is triggered. Until the last order date, the equipment manufacturer has time to decide upon a response. Depending on the part supplier, this delay may be from six months to one year. During this period, the equipment manufacturer has a whole range of solutions available, as shown later in Table 7-3. After the last order date has passed, the options begin to shrink, until the only options available after the last ship date has passed are the most expensive ones.

7.3.2 Degree of Life Cycle Mismatch

The degree of life cycle mismatch represents the ratio of the useful service life of a product to the expected life of the parts used. For an avionics product this ratio may be five to ten, meaning that the useful service life of these products is five to ten times longer than the expected life of the parts used. For a down-hole well package used in oil exploration, the ratio may be very small, for in some applications these may have a life requirement of only a few hours. Thus, the effect of obsolescence is not always a problem; it depends on the service life of the product, whether it is meant to be repairable, and whether it is intended to be produced in further quantities over a period of time.

7.3.3 Number of Products Using the Obsolete Part

Many parts are used in multiple products, and so the effect of an obsolescence event can vary widely in severity. The more products in which a part is used, the more effort and cost is involved in resolving the situation. Thus, the number of distinct products a part is used in affects the choice of corrective action. The objective, as always, is to find a solution that minimizes the overall cost. This is a complicated optimization problem that has many solutions, depending on the timescale over which one wishes to minimize the overall cost. The solution to the problem is subject to many unknowns, such as future market, failure rate, and the arrival of competing products.

7.3.4 Volume Requirement

The future demand for a part is governed by the planned production run, expected future orders, and expected future requirements for product support. The first of these is easily established from the factory schedule, but the latter two must be estimated. In addition, a part is usually used in more than one product, so forecasting has to take into account all of these products and their variants.

The two major factors are discussed next, but whatever numbers are finally decided upon, an adequate means of stock control is then needed to ensure that the allocations made for one purpose are not eroded by inaccuracies or changes in the other. For example, an organization that is contractually obliged to provide spare parts to third-party repair houses on demand should ensure that the contract provides for the third-party repair house to be reciprocally required to provide full identification of the item under repair. This prevents precious stocks of irreplaceable spare parts from being used on unrelated products.

7.3.5 Product Support

The need for parts to support products through to the end of life is particularly difficult to estimate. Useful data are available for repair operations if records have been kept on the types and numbers of parts failing in service. In many cases, the repair operation will be a third-party shop that has little incentive to collect this data. These records are notoriously difficult to maintain, and their accuracy is often suspect, as there is usually no time or resources spent on confirming that a removed part is actually faulty or is the root cause of the failure. Of course, where the product is not intended to be repaired, such as in the down-hole well-drilling application mentioned previously, or in space satellites, then the product support requirement does not arise.

7.3.6 Number of Obsolete Parts in a System

Obsolescence occurs at unexpected times and can lead to continuous scrambling to keep pace. One of the options available is to redesign the product to remove obsolete parts. This is an expensive process that has implications for the support of existing products in service, so this option is frequently not the first choice. Careful management and judicious use of bridge buys can put the manufacturer in a position to resolve many obsolete parts in a single redesign. If the redesign also provides some enhancement to the equipment, such as added functionality, reduced cost of production, or improved reliability, then there may be additional benefits.

7.3.7 Future Market

The estimation of future demand is usually based on previous market research data. A complicating factor is derivative products that may bear a strong resemblance to a current product in terms of the parts used, but that carry a different part number or product name.

7.3.8 Turnaround Time Available for Resolution

A major part of the cost of managing obsolescence arises from the time and effort required to continually monitor the entire portfolio of parts used in order to detect when a part is about to be made obsolete or has already become obsolete. Those companies that rely on the placement of an order for parts and face the subsequent rejection of them by the supplier will find that their reaction time is drastically curtailed. The first important step is to establish the processes needed to ascertain part discontinuance status on a continuous basis and to trigger an action based on a change of that status from active to obsolete.

7.3.9 Requalification Requirements

In most industries, the substitution of a new part may require that the product be requalified. This occurs in avionics when a subset of the original qualification tests needs to be repeated. The need for requalification arises partly from regulatory requirements, as in civil avionics; but, in general, requalification is an essential step to provide confidence in the changed design. Where qualification by similarity cannot be invoked, these tests are often quite expensive, time consuming, and have an obvious impact on the solution choice.

7.4 REACTIVE OBSOLESCENCE MANAGEMENT CHECKLIST

In Table 7-3 the reactive resolution options described previously are listed, along with the factors that need to be considered when choosing the most cost-effective strategy. In some cases, a combination of more than one strategy can be used to solve an obsolescence problem (such as bridge buy and a redesign).

7.5 REACTIVE OBSOLESCENCE MANAGEMENT GUIDELINE

Figure 7-6 shows the reactive obsolescence management guideline regarding how to implement a reactive obsolescence management system within an organization affected by obsolete parts.

Strategy	End of Life Notice			Degree of Mismatch		Number of Unique Products Using the Obsolete Part		Total Forecast Volume of Obsolete Part		Number of Different Obsolete Parts in Systems Affected in Each Product		Future Market for the Product		Potential for Producibility Enhancement		Turnaround Time Available for Resolution (Weeks)			Re-qualification Required	
X – potential solutions	Expired	Issued	Not Issued	Low	High	Low	High	Low	High	Low	High	Low	High	Low	High	< 20	20-60	> 60	Yes	No
Negotiation with the Manufacturer		X		X****			X**		X	X			X	X				X		X
Existing Stock		X		X****		X*	X**	X		X		X	X	X			X			X
Reclamation	X				X****	X*		X		X		X		X			X		X***	
Alternate Part	X	X	X	X****	X****	X*		X	X	X		X	X		X	X			X***	
Part Substitution	X	X	X	X****	X****	X*		X	X	X		X	X		X	X			X***	
Uprating	X	X		X****	X****	X*	X**	X		X		X		X		X	X		X***	
Aftermarket Sources	X			X****	X****	X*	X**	X	X	X		X	X	X			X		X***	
Emulation	X				X****	X*			X	X			X		X		X		X***	
Redesign	X				X****		X**		X		X		X		X			X	X	
Reverse-Engineering	X				X****	X*	X**		X	X			X		X			X	X	
Lifetime Buy		X	X	X****		X*	X**	X		X			X	X			X			X
Bridge Buy		X		X****		X		X		X		X		X		X				X

* Easier to forecast needed quantity but more risk of having "excess" parts.
** Spreads out the risk, but harder to forecast needed quantity.
*** Depends on part form-fit-function and customer/regulatory requirements.
**** Depends on equipment supplier's objectives and end product.

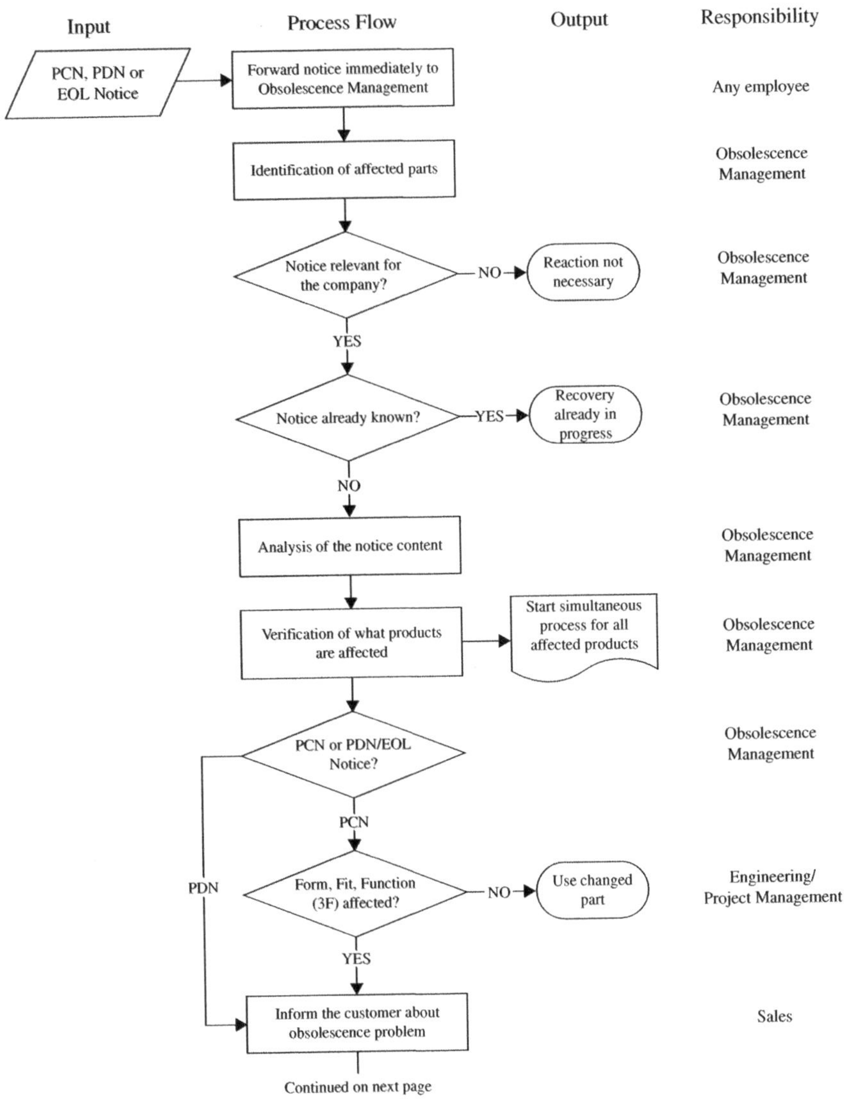

FIGURE 7-6 Reactive obsolescence management guideline.

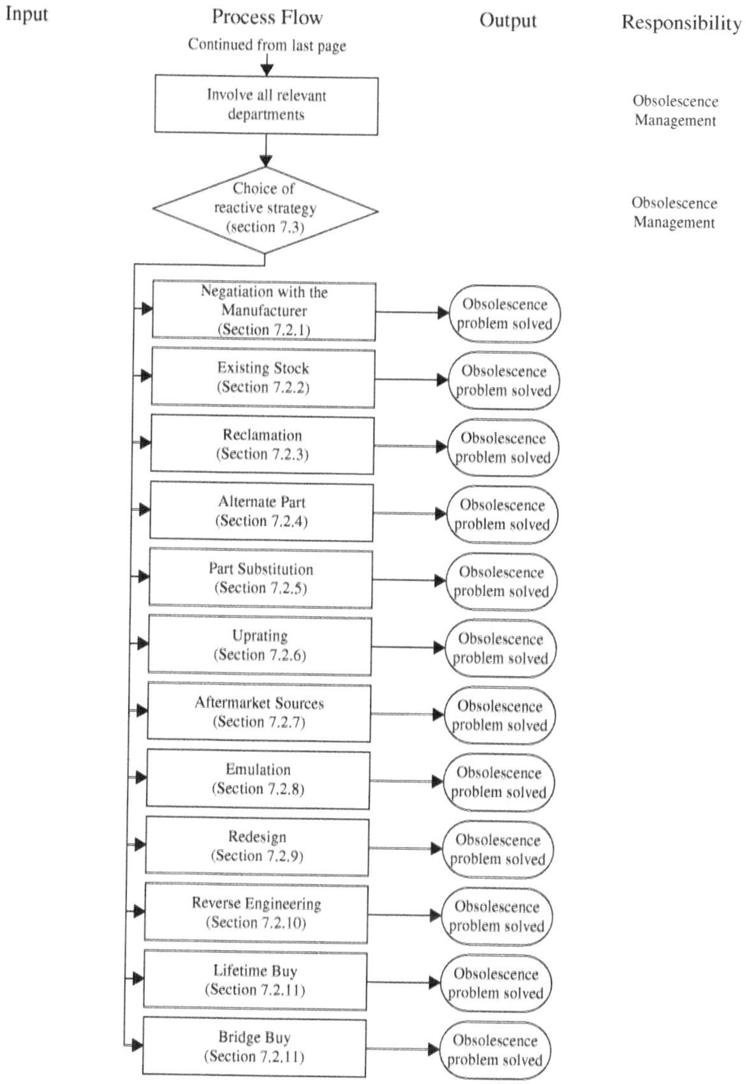

FIGURE 7-6 Continued

Chapter **8**

Proactive Obsolescence Management

This chapter describes proactive obsolescence management. The idea is to proactively track life cycle information on selected parts in order to prevent obsolescence-driven risks such as production stops and expensive redesigns.

Proactive obsolescence management means that the availability of parts is being monitored and actions to manage obsolescence are being taken prior to a part's actual discontinuance. Obviously, most organizations will not have enough resources to proactively manage all the parts in their systems, and therefore parts need to be ranked in order of importance. The criteria for proactive management is generally as follows:

- There has to actually be some demand for the part (if you are never going to need the part again, proactive management is not necessary).
- The part has to be at some risk of going obsolete.
- There has to be some risk of running out of the part if it does go obsolete (if five times as many parts as will ever be needed are currently in stock, proactive management is not needed).
- The part has to be in the critical path and pose some level of difficulty to manage if it goes obsolete.

As resources are available, proactive management organizations open DMSMS management cases for the non-obsolete parts starting with the highest priority part. Ideally, one should start proactive resolutions as soon as possible,

Strategies to the Prediction, Mitigation and Management of Product Obsolescence, First Edition. Bjoern Bartels, Ulrich Ermel, Michael Pecht and Peter Sandborn.

but proactive obsolescence management helps in all stages of the life cycle of a product.

8.1 MEMBERS OF THE PROACTIVE OBSOLESCENCE MANAGEMENT BOARD

It is advisable to include project participants from different departments in the proactive management process. For an electronic assembly board, the product manager inherits the most important position. If it exists, the obsolescence management department, which will provide relevant data throughout the product's life cycle, will also play a central role.

For realistic product pricing of an electronic assembly board, product management and sales should participate. In most cases, both departments together make decisions on future reactive activities and approve the later required budgets. The development department's participation is essential because of their expert knowledge on technical realization. The purchasing department contributes know-how on parts' prices, availabilities, and lead times and can evaluate the credibility of a supply's long-term delivery programs. Supply-chain management issues can either be accounted for by the supply-chain management department or the strategic purchasing department. The service department is an option and depends on the level of aftersales service provided for the product. Under certain circumstances (such as strong customer focus) the customer might also participate in the process.

8.2 SCHEDULE AND MILESTONES

During the product definition phase, the members of the Proactive Obsolescence Management Board should define the projected life cycle time and planned shipments of the electronic assembly board as far as possible. Expertise from similar past projects should be included. If the projected life cycle time is less than two years or if a "one-shot" production is planned, proactive obsolescence management focuses on development, prototyping, first production run, and aftersales support. Otherwise, the project will be reviewed after a defined span of time.

In the next step, all the participants are requested to agree on the defined time span. The purpose of this step is to raise awareness and prevent making narrow-minded future decisions.

The milestones are as follows:

1. First component selection is finished.
2. Initial obsolescence risk analysis is performed on the entire bill of materials.
3. Create a prioritized list of part candidates for proactive management.

4. Select parts from the proactive management list and open resolution cases for them as resources allow.
5. Monitor the availability of all the parts.
6. When parts that are not proactively managed are discontinued, the reactive management process is initiated (cases opened, resolved, and closed).
7. Reprioritize the proactive management list (add or subtract parts from it) as needed and continue monitoring all parts.

8.3 INITIAL OBSOLESCENCE RISK ANALYSIS

Once the initial selection of required components has been completed, an obsolescence risk analysis is performed. The purpose of this analysis is to get an overview of future availability, possible alternatives, and critical single-source items. This analysis delivers relevant bill-of-material (BOM) data (including availability forecasts), identifies risky materials (such as single-source components), and provides an overall health status. These elements are described in the following sections.

The initial analysis is of major interest, because it prevents the design-in of already obsolete components and components that are close to obsolescence. Therefore, the results of this analysis should be considered before the final design is finished and qualification started.

There are several ways the obsolescence management department can gather information on the preselected components. A fast way to access this data for electronic piece parts is to subscribe to one of many commercial databases. An organization's individual requirements and product and part mix determine which service fits best to the project. Some services are specialized for military parts and others offer more general commercial part databases. Generally, services are online and thus require the upload of part lists over the Internet. Solutions with post-firewall databases exist as well; these services generally cost more. After uploading the parts list, the online tool indicates how many part numbers it recognizes. If parts are not recognized, part numbers may need to be "messaged," or a manual part number inquiry can be ordered separately. Finally, the part list is returned with the relevant information, as shown in the next section. Note, this discussion is specific to electronic piece parts (chips, passive parts, connectors, and so on); general database support for assemblies and software do not exist.

If no database access exists, or parts are not recognized by the database, then either the organization has to research the parts themselves or pay to have a service research them (see Chapters 4 and 5).

8.3.1 BOM Management

The preselected parts list and later the BOM are required to have the following data:

1. Manufacturer part number
2. Manufacturer name
3. General function (resistor, capacitor, memory, and so on)
4. Specific function (for memory, 128Mb NOR Flash, and so on)
5. Alternatives (if any exist)

8.3.2 Material Risk Index

The results of the initial analysis or monitoring of the parts' availability contain information on the current part status and the current life cycle stage of the part. Depending on the forecasting methodology, the forecast has to be performed by the obsolescence department (see Chapters 4 and 5), or the predicted availability is already provided by a database service. Moreover, the number of sources and alternatives is known.

Based upon these criteria, a specific material risk index is created. If the part status is not recommended for new designs or is already obsolete, the material risk index will flag this part as red. If the prognostic availability span does not match with the project requirements, then the part will be flagged orange. If the part is only purchasable from a single source, it will be flagged orange as well. If there is no information available, the part will be flagged blue. These parts require actions to get more information. These parts are as critical as obsolete parts.

All other remaining parts are flagged green.

8.3.3 Health Monitoring

A key measure of the health of a system and a measure of the degree to which proactive management is being used (and succeeding) is the progress indicator given by SD-22 (2010)

$$PI = 100\frac{(G + Y_1)}{(G + Y_1 + R + Y_2 + B)}$$

where

G = two or more suppliers
Y_1 = one supplier and funded solution
Y_2 = one supplier and no funded solution
R = obsolete part and no solution
B = unknown status

The *PI* simply measures the fraction of parts that are "under control." One can modify the *PI* based on whatever criteria make sense to one's organization. Generally, scores of 90–100 indicate a DMSMS management best practice program; 80–89 indicates a solid DMSMS management program; 70–79 indicates that one probably has not figured out how to research the parts of unknown status yet; 55–69 indicates that one is starting to actively engage the available resources; below 55 indicates some potentially catastrophic problems.

PI can be calculated for a specific product or system, and it provides a measure of the supportability of the product or system. *PI* can also be applied to a whole database of active parts, in which case it provides a measure of the maturity of a DMSMS management process.

8.4 TRACKING PARTS' AVAILABILITY

Proactive obsolescence management relies on the material risk index staying current. Therefore, it is reasonable to perform the initial analysis on a regular basis. The initial analysis is the same as monitoring the parts' availability. The time span between these analyses depends on the health monitoring. If a BOM or preselected part list has numerous critical parts, the time span should be reduced. For instance, if the BOM contains several memory parts, it may be reasonable to check the BOM every two months. If the BOM consists only of passive components, the monitoring of parts' availability could be performed every six months (or longer). The frequency of analysis is not necessarily a constant; it may be set by a specific project's requirements.

8.5 PRODUCT OBSOLESCENCE AND AFTERSALES

All products reach discontinuance as some point in time. Once the proactive obsolescence management board makes the decision to obsolete an electronic assembly, there are certain rules an organization should follow. Customer notification is a clear must. For the correct on-time notifications, please refer to Chapter 2. For a quick reference, customers should be informed about one year in advance. It depends upon the product, but it might be reasonable to change the part number in order to avoid future confusion. Obsolescence does not necessarily mean that there will be no future demand. Therefore, all technical data should be stored in a safe place. Maybe the service department will need access to the data sometime in the future. Another possibility is to sell the document to customers or other organizations that could use it for system sustainment. Some semiconductor manufacturers sell off everything related to an obsolete part. The aftermarket suppliers may purchase the information from the semiconductor manufacturers in some cases.

Chapter 9

Strategic Obsolescence Management

To be cost effective, an obsolescence management plan must expand its coverage to include reactive, proactive, and strategic elements. The key methods involve minimizing the impact of component obsolescence by:

- Considering obsolescence management options during the part selection process
- Instituting proactive information sharing and standardization systems for improved predictive capability
- Developing strategic methods to define, design, partition, acquire, and use (maintain) products
- Maximizing availability of components by identifying, using, and supporting all available resources for procurement of components that meet the application requirements
- Thinking beyond single parts and components and applying management above the piece-part level
- Planning the life cycle management of the systems, that is, determining the optimum mix of reactive mitigation and design refreshes to manage obsolescence

A prerequisite for successful obsolescence management is the development of a plan and processes guide that details the infrastructure and processes to be followed in product design, manufacturing, and support.

Strategies to the Prediction, Mitigation and Management of Product Obsolescence, First Edition. Bjoern Bartels, Ulrich Ermel, Michael Pecht and Peter Sandborn.

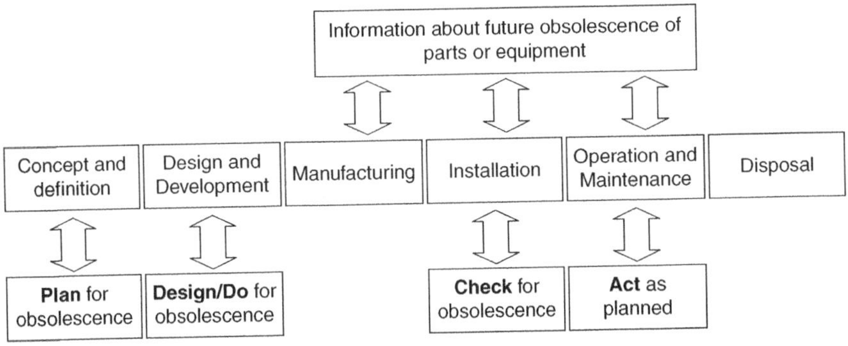

FIGURE 9-1 Strategic obsolescence management during product life cycle (adapted from IEC-62402, 2004).

Figure 9-1 shows that implementing a comprehensive obsolescence management plan is possible during all life cycle phases of a product or system. However, it is most effective if obsolescence management begins in the product development process during the definition and conceptual design of the product (Tomczykowski et al., 2000).

The most effective way to mitigate the risks and costs of obsolescence is to accommodate rapid component changes by being able to change product designs and production processes quickly at all levels (Condra et al., 1997; Condra and Matthews, 1999). If such a design and manufacturing environment can be achieved, managing component obsolescence will be like any other volitional design modification. The goal of the parts selection and management team in managing obsolescence must be to create this condition. However, achieving this goal requires long-term planning and cooperation at all levels of the intra-organizational and external supply chain.

9.1 APPLYING PROJECT MANAGEMENT PRINCIPLES TO OBSOLESCENCE MANAGEMENT

The reactive obsolescence mitigation approaches described in Chapter 7 are methods whereby an unplanned obsolescence situation can be handled. Reactive methods may become very expensive over the lifetime of a system, may require time and resources to implement that are not available when needed, or may fail to provide the parts support required throughout the life of the system. Therefore, it is important to have strategic obsolescence management solutions at the design stage and over the entire life cycle of the product.

The problem of how to implement a successful, cost-effective, and comprehensive strategic obsolescence management plan persists. Every occurrence of obsolescence, as well as every project, bears the risk of economic failure and also additional costs and expenses; therefore, it is important to implement an appropriate management system to analyze the possible causes of obsolescence, to implement reasonable methods to minimize the risk of obsolescence, and to plan far enough into the future to maintain cost control (Horvarth, 2006).

The following sections describe how to implement a comprehensive strategic obsolescence management plan. The strategic procedures and tactics are based on the fundamentals of project management, and both benefits and limitations are addressed. The sections and; likewise, the strategies, are sorted in order of the typical development stages of a project, as shown in Figure 9-2.

The concept of project management is applied to strategic obsolescence management in the following manner:

- Phase 1: The initiation stage describes problem recognition and analysis (Olfert, 2008). This stage is used to analyze the current situation and to build a consistent knowledge base regarding obsolescence.
- Phase 2: The planning and design stage describes the assessment and decisions concerning the project's topic (Olfert, 2008). This stage is used to develop draft obsolescence management plans and detect the weaknesses, risks, and causes of obsolescence.
- Phase 3: The execution stage that describes the management, handling, and execution of a project (Olfert, 2008). This stage is used to strategically operate, perform, and lead the obsolescence management system.
- Phase 4: The monitoring and controlling stage is usually called the competition stage, and it describes problem solving and the final control of a project (Olfert, 2008). This stage is used to define, analyze, and evaluate the costs of obsolescence management.

Strategic obsolescence management guidelines are provided at the end of this chapter (Section 9.6) to show how to implement a strategic obsolescence management system within an organization. All insights of this chapter are subsumed in the guidelines at the end of the chapter.

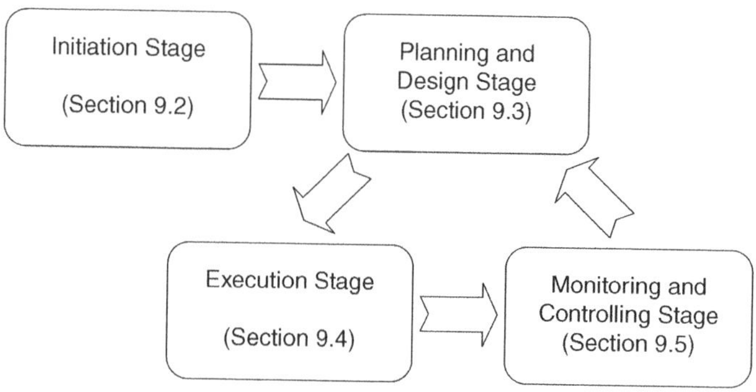

FIGURE 9-2 Typical stages of project management applied to strategic obsolescence management.

9.2 INITIATION STAGE

The initiation stage describes problem recognition and analysis (Olfert, 2008). Applied to comprehensive strategic obsolescence management, this stage is used to analyze the current situation and to build a consistent knowledge base regarding obsolescence, as described in the following sections.

9.2.1 Auditing

Auditing generally refers to inspection personnel using checklists to check and verify certain conditions within an organization. This is done through quality management or through obsolescence management (EN ISO 9000, 2000).

An audit based on quality is a systematic and autonomous analysis to detect whether operations and their results are in conformity with directives or standards. Furthermore, the directives are analyzed to determine whether they can be realized in an effective way and therefore are appropriate for reaching certain goals (Geiger, 1994).

Organizations should regularly be analyzed and evaluated regarding obsolescence management. Completed checklists provide an opportunity to analyze the responses and formulate an early assessment as to the potential depth and complexity of future obsolescence management issues. Furthermore, this activity provides a means for generating programmatic documentation that raises the awareness level of potential obsolescence impacts (Tomczykowski et al., 2000).

For example, a supportability checklist for new or proposed R&D programs is provided by the Air Force Research Laboratory Research and Development (R&D) Supportability Design Guide (AFRL, 1998). Many of the topics contained in this checklist not only ensure the supportability of new programs if the answers are affirmative, but they are also concerned with techniques to mitigate the impact of obsolescence (AFRL, 1998).

Potential interviewees for an audit are:

- Team members responsible for ensuring the supportability of the product for its intended life
- Component engineers
- Reliability engineers associated with qualification (and certification) of the system
- Team members who decide which obsolescence strategy to adopt
- Purchasing professionals who evaluate the financial implications of adopting a particular obsolescence management strategy

Table 9-1 shows an example of an audit questionnaire that presents some possible questions for an obsolescence management audit.

TABLE 9-1 Example Audit Questionnaire for Obsolescence Management

Audit Questionnaire		
No.	Question	Result
01	Do you deal with obsolescence or the negative effects of obsolescence on your organization?	-
02	What are the reasons for supply shortfalls caused by obsolescence within your organization and at suppliers?	-
03	Is a comprehensive obsolescence management system implemented in your organization?	-
04	Is your organization in possession of a general or program-specific obsolescence management plan?	-
05	Is this obsolescence management plan improved continually to ensure a constant qualitative performance increase?	-
06	Is your organization a member of an obsolescence management organization or do you participate regularly in industry and government obsolescence conferences and/or working groups?	-
07	What kinds of components are used for production (electronics, mechanics, textiles, and software)?	-
08	Is your organization affected by obsolescence caused by directives, rules, and other legislation imposed by governments (e.g., directives for environmental protection)?	-
09	Are you able to estimate the cost impacts of obsolescence on your organization?	-
10	Who is in charge of obsolescence management in your organization?	-
11	What obsolescence recovery tactics are used in your organization to react to an obsolescence problem?	-
12	Are you in possession of information regarding the life cycle stages of all parts and components used for production and/or maintenance?	-
13	Does your organization use an obsolescence health monitoring metric or tool?	-
14	Do you deal with the benefits and limitations of long-term storage?	-
15	Do you have an issue with counterfeit components and replicas?	-
16	Are effective anti-counterfeiting tactics implemented in your organization?	-
17	Do you have problems receiving PCN or PDN notices or with their content?	-
18	Are you in possession of reactive obsolescence management guidelines?	-
19	Is your organization regularly analyzed and evaluated regarding obsolescence problems through audits?	-
20	Is your entire product team educated to raise their awareness and avoid obsolescence problems?	-
21	Do you use product-specific process analysis for obsolescence problems?	-
22	Does your organization consider obsolescence management when you design your products?	-
23	Is design refresh planning optimization implemented in your organization?	-

(*Continued*)

TABLE 9-1 Continued

Audit Questionnaire		
No.	Question	Result
24	Do you use a preferred parts list?	-
25	Is an effective supplier management plan implemented in your organization?	-
26	Is obsolescence covered in contracts with suppliers and customers?	-
27	Do you monitor and control your obsolescence management system?	-
28	Are you in possession of strategic obsolescence management guidelines?	-
29	Are you able to estimate the cost and benefits of your obsolescence management system (e.g., return on investment)?	-
30	Is there a need for action to improve your obsolescence management system?	-

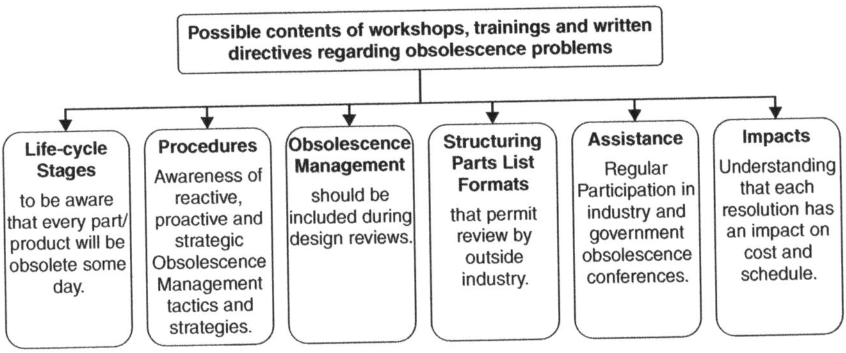

FIGURE 9-3 Possible contents of educational activities and methods for raising awareness.

9.2.2 Raising Awareness

One of the most effective tactics to avoid obsolescence problems is educating product teams to raise their awareness of the problem. Anyone bringing together product stakeholders from management, engineers, designers, buyers, and planners with those responsible for sustainment needs to understand that obsolescence is not only a significant industrywide problem but also a growing one (Tomczykowski et al., 2000).

Employees can be educated in workshops, training seminars, and also through written directives. Possible course contents and methods of raising awareness are shown in Figure 9-3.

Creating mailing lists and sending messages to groups of people or creating databases containing information about obsolescence problems will provide help. These tools can be used to enlarge the awareness of employees and improve the flow of important information.

9.3 PLANNING AND DESIGN STAGE

The planning and design stage describes the assessment and decisions concerning the topic of a project (Olfert, 2008). Applied to a comprehensive strategic obsolescence management, this stage is used to develop draft obsolescence management plans and detect weaknesses, risks, and causes of obsolescence, as described in the following sections.

9.3.1 Design Products to Avoid Obsolescence

Product designers must be knowledgeable about the electronic components industry, the forces acting on it, and the trends in its response to those forces. Strategically, this involves education about the industry and the supply chain upon which it feeds, rather than training in specific components and component designs. In the product design process, manufacturers should develop and use processes to consider future expected component changes, including obsolescence, in their internal design approval processes (preliminary design review [PDR], critical design review [CDR], and so on).

Equipment manufacturers use the available knowledge of the forces acting on them to estimate the amount by which parameters such as feature size or supply voltages will decrease, or the speed will increase, and so on. These trends can be accommodated in the designs by partitioning the design to ensure, if possible, that high-risk items are contained on replaceable modules within the products known as "shop discardable units" (SDUs) or replacement SDUs. These units are form, fit, and function replacements for the original module and can be substituted with minimal effects on the remainder of the LRUs.

Equipment manufacturers should develop plans for their products with regard to expected component changes over the entire life cycle of a product. Obviously, equipment manufacturers cannot predict exactly when or how component designs will change or when they will become obsolete. The picture becomes especially hazy when trying to look beyond the first 5 to 10 years of product life.

The equipment manufacturer is responsible for the entire life cycle, and if a life cycle plan for the product is developed, the equipment manufacturer can at least begin to understand what the right questions are. Each equipment manufacturer should develop a general plan for dealing with component obsolescence. In addition, there should be a specific plan, to be reviewed in PDR, CDR, and so on for the product. The plan might include, for example, a list of key components with a severe obsolescence impact; plans for last-time buys of key components; scheduled redesigns and design refreshes; and the impact on, and recommendations for, modifications to product operation, maintenance, and support, should particular contingency plans be implemented.

Generally, process analyses help identify the effects that influence the occurrence of obsolescence, thereby influencing the development of obsolescence management plans.

9.3.2 Process Analyses

Many process analyses are available to upgrade and enhance the quality and supportability of products and processes with the help of rating and test indices. The following sections describe different types of process analyses that are applicable to obsolescence problems. A comprehensive obsolescence management plan should also address process analyses to mitigate the risk of obsolescence (Sondermann, 1994).

9.3.2.1 Ishikawa/Fishbone Diagram The Ishikawa diagram, also referred to as a fishbone diagram or cause-and-effect diagram, shows the causes of a certain event. For obsolescence management, it shows the possible causes of the undesirable event of obsolescence (Geiger et al., 2008).

The systematic development of an Ishikawa diagram is usually done in brainstorming sessions.

Existing standards provide the main categories to trace back the cause of a certain event to its root causes (DIN ISO 9004, 2000). Below, these categories, also called Ms, are translated according to their meaning and extended with examples of how their meanings can be applied to an obsolescence event:

- Milieu (e.g., Suppliers and customers)
- Method (e.g., Methods of solving obsolescence problems)
- Machine (e.g., Health-monitoring tools)
- Man Power (e.g., Employees in an organization)
- Material (e.g., Raw, parts, components)
- Management (e.g., Obsolescence management)
- Money (e.g., Money lost due to obsolescence)

After sorting out the causes of a particular event, the causes are then classified into main and adventitious causes.

Figure 9-4 shows an example Ishikawa diagram applied to an obsolescence event.

9.3.2.2 Fault Tree Analysis (FTA) A fault tree analysis (FTA) is a deductive, top-down failure analysis. An undesired state of a system, in this case, the obsolescence of a part or component, is analyzed using Boolean logic to combine a series of lower-level events. FTA uses pathways within a system that can lead to an undesirable event shown in a graphic model. The pathways interconnect contributory events and conditions using standard logic symbols (AND, OR, and so on) to analyze system failure (Geiger, 1994; Geiger et al., 2008).

The basic constructs in FTA are causes, gates, and events. Different causes that generate an event are combined with gates. If the occurrence of either cause makes the top event occur, then these causes are connected using an

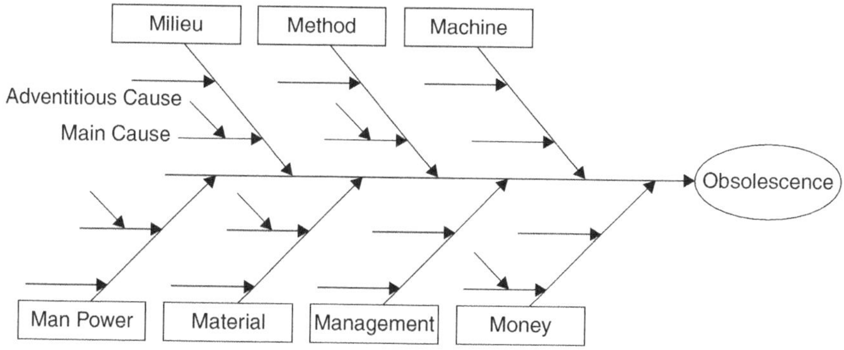

FIGURE 9-4 Ishikawa diagram for obsolescence event (adapted from DIN ISO 9004, 2000).

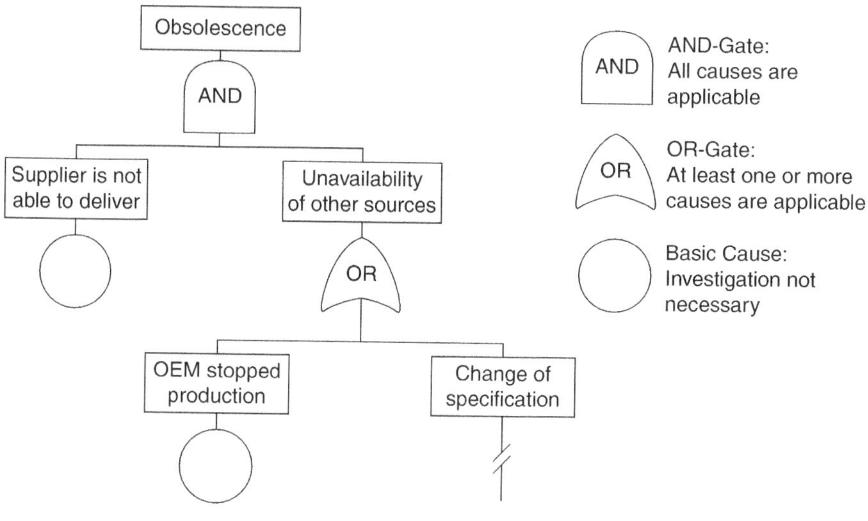

FIGURE 9-5 Fault tree analysis of obsolescence event (adapted from EN 61025, 2007).

OR-gate. If all causes need to occur to cause the top event, they are connected by an AND-gate (Geiger, 1994; Geiger et al., 2008). The output is a logical diagram that shows all causes that can lead to an undesirable event, in this case obsolescence.

Figure 9-5 shows an example FTA applied to an obsolescence event displaying different possible causes combined by gates.

Finally, FTA methodology is described in several industry and government standards—for example, NRC NUREG–0492 for the nuclear power industry, an aerospace-oriented revision to NUREG–0492 for use by NASA, SAE ARP4761 for civil aerospace, MIL–HDBK–338 for military systems, IEC standard 61025 for cross-industry use—and it has been adopted as European Norm EN 61025.

9.3.2.3 Failure Modes and Effects Analysis (FMEA) A failure modes and effects analysis (FMEA) evaluates the ways that equipment can fail or be improperly operated, and the effects these failures can have. An FMEA can also be applied to the problem of obsolescence occurrence, where the failure to sustain an item (piece of equipment) is caused by a particular obsolete part or part family. In an FMEA, each individual failure is considered to be an independent occurrence with no relation to other failures in the system. In short, FMEA is a methodology to analyze and discover all potential failure modes of a system, the effects these failures have on the system, and how to correct or mitigate the failures or effects on the system (Geiger, 1994). The general FMEA preparation process is shown in Figure 9-6.

As the correction and mitigation are usually based on a ranking of the severity and probability of the failure, a risk priority number (RPN) needs to be calculated. The RPN per failure is calculated by multiplying the failure severity by the occurrence probability and detection probability. Figure 9-7 shows rating criteria and classification and an example of a form that could be used for obsolescence management.

After calculating the RPN, corrective actions have to be recommended and applied for high-risk items or functions. High-risk items or functions have an RPN that is 125 or higher on a single rating scale for severity, occurrence, or detection of 8 or higher (Kersten, 1994).

The FMEA methodology is described in several standards, for example, MIL-STD-1629A. Furthermore, some organizations, such as the German Association of the Automotive Industry (VDA), provide standardized forms for the development of an FMEA.

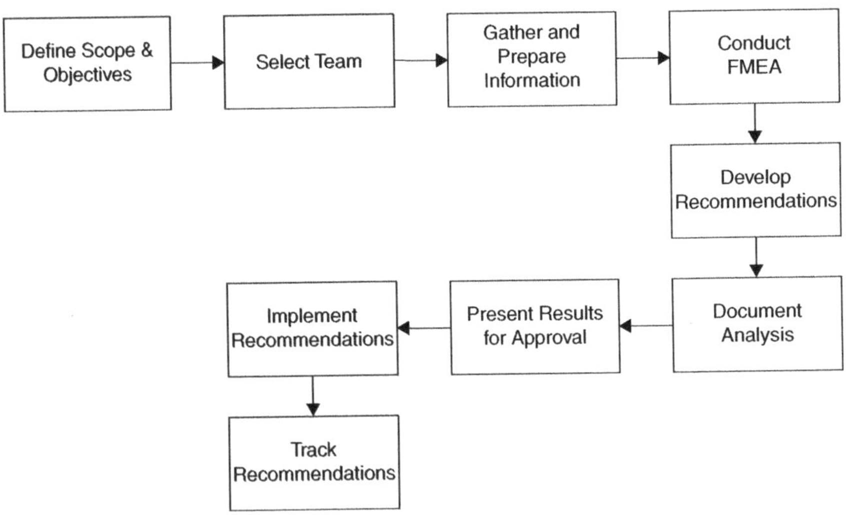

FIGURE 9-6 FMEA preparation process (adapted from FMEA Infocentre, 2010).

Severity Rating Scale # Description	Occurrence Rating Scale # Description	Detection Rating Scale # Description
1 None	1 Remote: Failure is unlikely	1 Almost Certain
2 Very Minor	2 Low: Relatively few failures	2 Very High
3 Minor	3 Low: Relatively few failures	3 High
4 Very Low	4 Moderate: Occasional failures	4 Moderately High
5 Low	5 Moderate: Occasional failures	5 Moderate
6 Moderate	6 Moderate: Occasional failures	6 Low
7 High	7 High: Frequent failures	7 Very Low
8 Very High	8 High: Frequent failures	8 Remote
9 Hazardous with warning	9 Very High: Persistent failures	9 Very Remote
10 Hazardous without warning	10 Very High: Persistent failures	10 Absolute Uncertainty

Item / Function	Potential Failure Mode	Potential Effect(s) of Failure	Sev	Class	Potential Cause(s)/Mechanism(s) of Failure	Occur	Current Design Controls	Detec	RPN	Recommended Action(s)

RPN = Severity × Occurrence × Detection

FIGURE 9-7 FMEA rating criteria and classification (adapted from ReliaSoft, 2010).

9.4 EXECUTION STAGE

The execution stage describes the management, handling, and execution of a project (Olfert, 2008). Applied to a comprehensive strategic obsolescence management plan, this stage is used to strategically operate, perform, and lead the obsolescence management system, as described in the following sections.

9.4.1 Forecasting the Product Life Cycle

Comprehensive obsolescence management planning should start during the development and conception phase of a new product.

System designers must consider the system design life while planning the life requirements of their own products, as their own product life cycle will influence the choice of an obsolescence avoidance strategy and also the life cycle management plans, including the scheduling of redesigns and refreshes of the system.

The following essential concepts should be included in the forecasting of a product's production and support life cycle (Tomczykowski et al., 2000):

- A minimum acceptable life determined by the threshold for earning a profit by selling and/or supporting the product or system. At least all development and manufacturing costs should be covered.
- The minimum acceptable operational life that is necessary to provide an operational capability that will satisfy the end user's and mission's needs.
- Consideration must also be given to the possibility of extension of the actual system life expectancy. An example is the B-52 aircraft, which is now many years beyond its objective life expectancy.
- Design refresh planning optimization should be addressed in the forecasting of the product life cycle, as it provides the advantage of treating multiple existing and anticipated obsolescence problems concurrently (see Section 9.4.9).

The goal is to optimally design a system or product to minimize the cost of managing inevitable obsolescence problems. Even if only some information is available due to uncertainties, the estimation of lifetime sustainment costs via life cycle planning should be possible to have an opportunity for sustainment cost avoidance (Singh, et al., 2002).

9.4.2 Parts Selection Process

It is important that designers have easy and direct access to obsolescence forecasting information related to the components that they are choosing. Equipment manufacturers may also find it beneficial to maintain a preferred parts list (PPL). The emphasis here is on preferred or recommended parts, since such lists are inherently difficult to keep up to date and a mandatory list would not be consistent with evolving product performance requirements.

New designs should be required to select components that can pass an internal selection process gate. The process should ensure that components are qualified for the application, from a vendor in whom the purchaser has confidence, and can be produced with adequate ongoing quality. There is an incentive to use components that have already demonstrated these attributes since the approval (qualification) costs are lower, but there should be no absolute requirement to do so. Technological progress leading to the proliferation of new parts will undoubtedly continue. Ideally, designers need to consider the total cost of ownership of their part selection decisions, that is, the costs of introducing a new component into their organization, qualifying the component and the manufacturer of the component, maintaining the component within their databases and qualification system, and inventory and support costs throughout the product life cycle (Prabhakar and Sandborn, 2011).

There is also a loss of consolidation benefits when opportunities for selecting existing approved components are not taken. Addressing this has the beneficial effect of reducing component selection costs and at the same time enlarging the market for those that remain. Taken across the industry, that translates into an enlarged market for the component supplier and reduces the chance of that market being insufficient to sustain the supplier's interest. Part selection and management plans should take into account the market size for each part. Such a policy is only partially effective, because a step technology capability increase can wipe out a market in a very short time as the volume users switch en-masse to the newer components.

Only when the service life of a product ends before the end of procurement life for all its components is there no risk of obsolescence associated with the product. The goal of system design teams should include the reconfigurability of subsystems of assemblies for easy modifications later in the system life cycle.

For electronic products, the following example solutions can be used:

- As redesigns can benefit from the use of programmable devices, the usage of programmable logic devices, such as field programmable gate arrays (FPGA) or programmable logic devices (PLD), is recommended.
- Flexible designs using alternative package styles/footprints (to account for transitions from one manufacturing process to another; for example, through-hole technology to surface mount technology) occur rapidly, which prevents extensive redesigns, even when using mature technology in new designs is unavoidable.
- Modular designs/upgrades at the subsystem level: Such designs can make module functional interfaces impervious to module changes, redesigns, substitutions, and upgrades. While the components used in individual modules are likely to become obsolete at the same rate as other

components, with proper controls, system integrators can preserve functions and interfaces. The modular design approach can minimize system redesign impact, and if modules are used across multiple products, some economies of scale may be obtained. Thus, by using modular designs, products and subassemblies can be made reusable, upgradeable, and scalable. To install upgraded versions of an existing component or package design, circuit card layout changes are common. Hence, high obsolescence risk components should be placed in subassemblies (line discardable units, LDUs) that should be discarded and replaced with alternative pin-for-pin compatible assemblies, even if the individual components change. PC manufacturers with microprocessors and memory cards do this routinely. An acceptable "threshold" cost, below which it is economical to discard a subassembly and therefore consider it an LDU, should be defined.

Typically, engineers should keep the following in perspective when implementing new designs:

1. Has the design been partitioned to simplify solutions to expected component obsolescence problems?
2. Have memory growth paths been incorporated into the design layout?
3. Do microprocessors have road maps defined by the manufacturer?
4. Does the design consider plans for future enhancements/upgrades?
5. Are pre-planned product upgrades identified?
6. Are planned design upgrades compatible with those of similar products using similar components?
7. Is design information archived in sufficient detail to facilitate future upgrades if the original design team is not available?
8. Are high-risk parts placed on subassemblies economical to discard?
9. Are cost estimates available for scheduled design upgrades?
10. Are support guarantees compatible with forecasting component availability?
11. Do minor variations in product operating procedures allow for the use of parts with a lower risk of obsolescence?
12. Have hardware dependencies on software been accounted for?

9.4.3 Demand Specification

The holistic demand for parts is composed by the sum of all bills of materials (BOMs) for different products. These BOMs list and identify all parts and components that are needed to produce or maintain a product or system.

To obtain an exact demand specification, it is necessary that no parts be missing from the BOM and that all parts be specified distinctly.

For clear identification of a part or component, the following minimum standard description should be used:

- Name and/or description
- NATO (or National) stock number (NSN) (for military use)
- Part number(s) from the OEM(s)/OCM(s)
- Name(s) of the OEM(s)/OCM(s)
- Commercial and government entity (CAGE) code of the OEM(s)/OCM(s) (for military use)
- Number of parts needed per board (or other level of indenture)
- Issue of the drawings or the software (if applicable)

Note, BOMs may contain more than one approved supplier and OEM/OCM part number for some parts since multiple parts are qualified for use.

A further advantage of clear and consistent specifications of used parts is that there is no need for manual correction of part numbers and manufacturers before uploading a BOM in a health-monitoring tool.

In addition to a clear description of parts, it is necessary to comprehensively specify the demand for parts and components. For this reason, not only do BOMs from serial production need to be observed, but also BOMs from design, development, and maintenance should be considered.

9.4.4 Supplier Management

Supplier management usually deals with the management of suppliers and their input, such as materials and services. The amount of money avoided by the organization is the commonly measured variable. However, in this case, the risk of obsolescence is one of the other critical aspects of supplier management. The goal is to keep the negative effects of obsolescence under control and to ensure that all relevant Product Change Notifications (PCNs) and Product Discontinuance Notifications (PDNs) reach the organization on time. The early and comprehensive integration of suppliers in the organization-owned obsolescence management plan ensures that needed information reaches the organization and there is prevention from obsolescence (Rogokowski, 2003).

The relevant portions of a successful supplier management plan are the selection, integration, and development of suppliers, as shown in Figure 9-8.

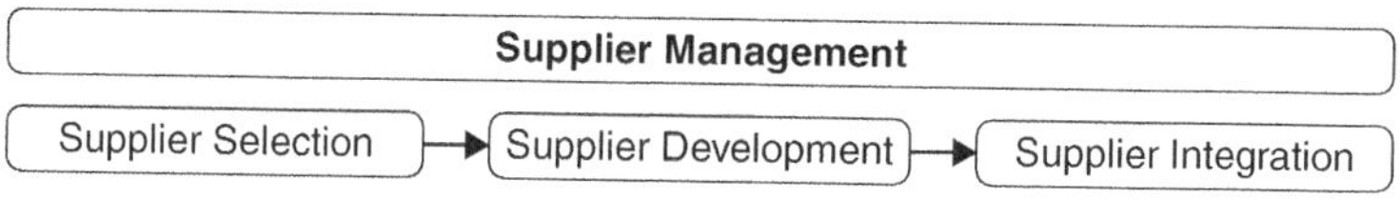

FIGURE 9-8 Supplier management.

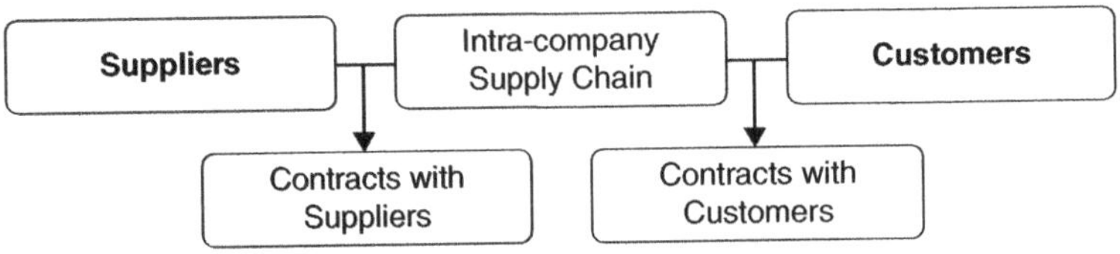

FIGURE 9-9 Contracts in the supply chain.

An appropriate selection of the source of supply is essential to maintain part availability. To evaluate suppliers with regard to obsolescence risk, the following methods can be helpful:

- Supplier questionnaire/supplier self assessment[1]
- Supplier audit[2]

These methods allow an organization to collect information about different suppliers and discern the strengths of suppliers and areas in which improvements can be made, which are then monitored for progress.

Supplier development refers to direct intervention in a supplier's organizational and operational structures. with the goal of raising performance and output (Muessigmann, 2007). For example, it might be helpful to develop and implement the obsolescence management system of a supplier to raise awareness about obsolescence and enhance the flow of information.

A supplier can develop its own ideas and strategies to solve and prevent obsolescence problems if the supplier is already integrated into the design and development of new products (Rogokowski, 2007).

It might be appropriate to include some issues concerning the previously mentioned portions of the supplier management plan in agreements or contracts with suppliers, as shown in the following section (Rogokowski, 2007).

9.4.5 Contractual Language

Contracts are needed on both sides of the supply chain with suppliers and customers. This section describes the necessity and the contents of contracts, as shown in Figure 9-9.

Contracts with suppliers and part manufacturers ensure that information from the original part manufacturer (such as PCNs and PDNs) reach the organization in a timely manner and ensure that an appropriate reaction to an obsolescence problem is taken. In addition, contracts mitigate the risk of

[1]Supplier questionnaires or supplier self-assessments are comprehensive, systematic, and regular reviews of an organization's activities to discern clearly its strengths and areas in which improvements can be made.

[2]Supplier audits are essential to verify the existence of objective evidence of processes and to assess how successfully processes have been implemented.

obsolescence and help evaluate the probability of the occurrence of obsolescence. Further, contractual policies can ensure that in the case of part obsolescence, the supplier has to take responsibility to solve a problem. Hence, the risk and also possible costs and effort can be shared with the supplier. When negotiating contracts with suppliers, it should be ensured that the supplier is able to meet the agreed requirements and responsibilities regarding obsolescence, as stated in the contract.

According to the "SD-22—Diminishing Manufacturing Sources and Material Shortages (DMSMS) Guidebook," published by the Defense Standardization Program Office (DSPO) (SD-22, 2010) the Naval Inventory Control Point (NAVICP) provides a standard set of contract clauses. In these clauses it is stated as an example that:

> The Contractor is responsible for managing obsolescence over the entire period of the contract, and notwithstanding any obsolescence issues or problems, the Contractor remains responsible for meeting all performance and other requirements of this contract. This obsolescence management responsibility includes an ongoing review and identification of actual and potential obsolescence issues, including but not limited to obsolescence of components, assemblies, sub-assemblies, piece parts, and material (hereafter referred to for purposes of this section only as "parts and/or material"). The Contractor is responsible for all costs associated with obtaining a replacement if and when any parts and/or material become obsolete. The costs for which the Contractor is responsible include, but are not limited to, the costs of investigating part availability, interchangeability and substitutability, locating part replacement, vendor interface, engineering efforts, testing requirements, internal drawing changes, etc. The Contractor shall prevent any additional costs from being incurred by the Government due to obsolescence. Any configuration changes due to obsolescence shall be approved in accordance with the Configuration Management requirements of this SOW. The contractor shall provide the Government with obsolescence status briefs, as part of the periodic program reviews provided for under the contract.

In addition, SD-22 (2010) provides another paragraph given from the NAVICP asking the contractor to provide a plan of what they propose to do to meet their responsibilities. In this clause it is stated that

> The Contractor shall develop and submit as part of its proposal (with an advance copy supplied to the Government at time of cost estimate submission), an Obsolescence and DMSMS Management Plan for managing the loss, or impending loss, of manufacturers or suppliers of parts and/or material required for performance of this contract. This plan will also address DMSMS Management. At a minimum, the plan shall address the following:
>
> - Means and approach for providing the Government with information regarding obsolescence and DMSMS issues:
> - Planned resolution of current obsolescence and DMSMS issues

- Parts list screening
- Parts list monitoring
- Processing Government Industry Data Exchange Program (GIDEP) (www.gidep.org) DMSMS Alerts
- Processing DLA DMSMS Alerts
- Communication with and availability of information to the government
- Means and approach for establishing obsolescence and DMSMS solutions
- Plan for conducting DMSMS predictions.

Furthermore, some part manufacturers are also willing to provide a procurement availability guarantee for certain parts. Contractors then ensure the maintenance of an inventory of items for future production use through contractual agreements. Furthermore, under some circumstances, a supplier may guarantee the long-term availability of a part or family of parts. These parts are called "high reliability" (HiRel) parts. These arrangements usually come along with higher costs of parts (U.S. DoD, 2005).

In contrast to contracts with suppliers, contracts with customers protect customers from assuming the whole responsibility of obsolescence. Contracts for long-term delivery should only be signed when additional costs for resolving obsolescence problems are excluded. However, that does not mean that customers' problems should not be solved. It is more about reacting to an obsolete part with regard to agreed monetary compensation. Existing contracts should be checked to see if such clauses regarding the sharing of the risk of obsolescence are currently included. In some cases, renegotiation is recommended.

9.4.6 Special Obsolescence Management Capabilities

The U.S. government and certain equipment manufacturers have their own capabilities to manufacture some components. They are capable of providing ongoing supplies of selected obsolete parts. The technical capabilities of these facilities usually support several generations of technology found in semiconductor manufacturing companies.

9.4.7 Streamlining Regulatory Procedures

For some low-volume industry sectors, one hurdle to using innovative design practices is inflexible certification and slow qualification. Both the certification and qualification of avionics products (or the equivalent military approval process) can occupy a high percentage of the development time; often, parts are already obsolete by the time a product is certified. Equipment manufacturers and regulatory agencies must work together to streamline the certification and recertification processes if they are to accommodate the new realities of the rapidly changing components industry. Design changes should be investigated at the subsystem level rather than at the assembly or the component levels. The key is to obtain confidence and, hence, certification credit for re-used hardware via service experience in commercial and industrial applications in lieu of extensive and expensive retesting.

9.4.8 Management above the Piece-Part Level

Management of obsolescence above the piece-part level needs to be addressed, as program managers are becoming more concerned about the timing of redesign solutions for whole systems.

For systems and products in current production, the best solution may be to procure enough parts to support continued production until a redesign can be made in the future (make bridge buys). Organizations need to look at the overall product line strategy affecting the future requirements for the hardware in question. Factors affecting the decision process include (Porter, 1998):

- Continued market for the product
- Manufacturability of the system (including the potential for producibility enhancement and the risk of manufacturing processes going obsolete)
- Risk of other obsolete parts in the system or product
- Other systems or products affected by this obsolete part
- Impact on users of a redesign, such as lack of spares commonality, costs associated with having a second system/product in their inventory, and retrofit costs
- Economics of the decision

For defining the best solution, break-even year charts provide quantitative guidance in developing a solution for component obsolescence. It is important to have tools that will provide an understanding of the economic factors involved in the determination of the optimum year for planning a redesign (Porter, 1998).

The following sections address problems, such as defining the best time for redesigning a system or product, and the usage of open systems and hardware-software independence to strategically manage obsolescence problems.

9.4.9 Design Refresh Planning Optimization

Because of the long manufacturing and field lives associated with sustainment-dominated systems, they are usually refreshed or redesigned one or more times during their lives to update functionality and manage obsolescence. Unlike high-volume commercial products in which redesign is driven by improvements in manufacturing, equipment, or technology, for sustainment-dominated systems, design refresh is often driven by technology obsolescence that would otherwise render a product unproducible or unsustainable.

Ideally, there should be a methodology that determines the best dates for design refreshes and the optimum mixture of actions to take at those design refreshes. The goal of refresh planning is to determine:

- When to design refresh (or the optimum frequency for design refresh)
- What obsolete system components should be replaced at a specific design refresh (versus continuing with some other obsolescence mitigation strategy)
- What non-obsolete system components should be replaced at a design refresh

Numerous research efforts have worked on the generation of suggestions for redesign in order to improve manufacturability. Redesign planning has also been addressed outside the manufacturing area, such as general strategic replacement modeling, re-engineering of software, capacity expansion, and equipment replacement strategies. All of this work represents redesign driven by improvements in manufacturing, equipment, or technology (strategies followed by cutting-edge products), not design refresh driven by technology obsolescence, which would render the product unproducible and un-sustainable. It should also be noted that manufacturers and customers of sustainment-dominated systems have as much interest in "design refresh" as "redesign" (Sandborn, et al., 2003).[3]

9.4.9.1 *Porter Model for Refresh Planning* The simplest model for performing life cycle planning associated with technology obsolescence (specifically, electronic part obsolescence) was developed by Porter (1998). Porter's approach focuses on calculating the net present value (NPV) of last-time (bridge) buys and design refreshes as a function of future date. As a design refresh is delayed, its NPV decreases and the quantity (and cost) of parts that must be purchased in the last-time buy to sustain the system until design refresh increases. Alternatively, if design refresh is scheduled relatively early, then the last-time buy cost is lower, but the NPV of the design refresh is higher. In the simplest form of a Porter model, the cost of the last time buy (C_{LTB}) is given by (Sandborn 2011)

$$C_{LTB} = \begin{cases} 0, \text{ when } i = 0 \text{ or if } Y_R = 0 \\ P_0 \sum_{i=1}^{Y_R} Q_i \quad \text{if } Y_R > 0 \end{cases} \tag{9-1}$$

where

i = years until design refresh

P_0 = price of the obsolete part in the year of the last-time buy (year 0 in this case)

[3]Refresh refers to changes that "have to be done" in order for system functionality to remain useable. Redesign or technology insertion means "want to be done" system changes, which include new technologies to accommodate system functional growth and new technologies to replace and improve the existing functionality of the system (Herald, 2000).

Y_R = year of the design refresh (0 = year of the last-time buy, 1 = year after the last-time buy, and so on)

Q_i = number of parts needed in year i

Equation 9-1 assumes that the part becomes obsolete at the beginning of year 0 and that the last-time buy is made at the beginning of year 0.

The design refresh cost for a refresh in year Y_R (in year 0 dollars), C_{DR}, is given by

$$C_{DR} = \frac{C_{DRI_0}}{(1+r)^{Y_R}} \tag{9-2}$$

where

C_{DRI_0} = design refresh cost in year 0

r = discount rate

The total cost for managing the obsolescence with a year Y_R refresh is given by

$$C_{Total} = C_{LTB} + C_{DR} \tag{9-3}$$

Figure 9-10 shows a simple example using the Porter model. In this case, C_{DRI_0} = \$100,000, r = 12%, Y_R = 20, Q_i = 500 (for all i from year 1 to 20; Q_i = 0 thereafter), and P_0 = \$10. In this simple example, the model suggests that the optimum design refresh point is in year 7.

The optimum refresh year from the Porter model can be solved for directly, assuming that the demand quantity is the same in every year ($Q = Q_i$ for all i = 1 to Y_R) (Sandborn, 2012).

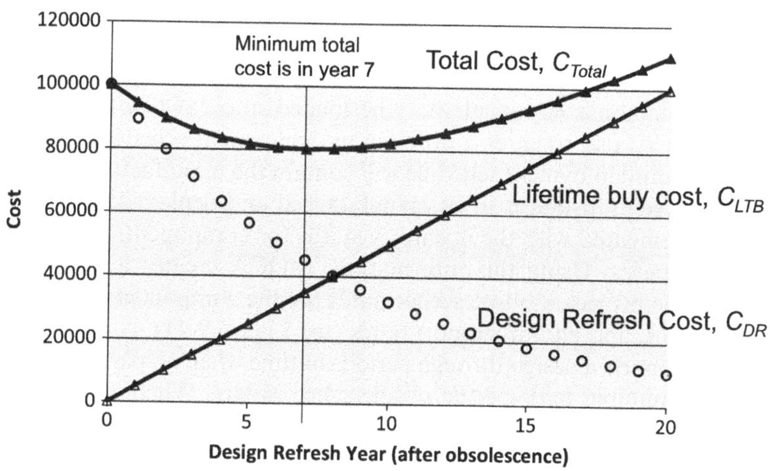

FIGURE 9-10 Example application of Porter's refresh cost model (Sandborn, 2012).

$$Y_R = \frac{1}{-r} \ln\left(\frac{P_0 Q}{r C_{DRI_0}}\right) \tag{9-4}$$

For the example given earlier, Equation 9-4 gives $Y_R = 7.3$.

The Porter model performs its tradeoff of last-time buy costs and design refresh costs on a part-by-part basis. Although the simple Porter approach can be extended to treat multiple parts, and a version of Porter's model has been used to plan refreshes in conjunction with lifetime buy quantity optimization in Cattani (2003), it only considers one single design refresh at a time. To treat multiple refreshes in a product's lifetime, Porter's analysis can be reapplied after a design refresh to predict the next design refresh, effectively optimizing each individual design refresh. However, the coupled effects of multiple design refreshes (coupling of decisions about multiple parts and coupling of multiple refreshes) in the lifetime of a product are not accounted for in the Porter model, which is a significant limitation of the Porter approach.

9.4.9.2 The MOCA Refresh Planning Model A more complete optimization approach to refresh planning called Mitigation of Obsolescence Cost Analysis (MOCA) has been developed that optimizes over multiple refreshes and multiple obsolescence mitigation approaches (the Porter model only considers last-time buys) (Singh and Sandborn, 2006). Using a detailed cost analysis model, the MOCA model determines the optimum mix of reactive mitigation and design refreshes during the field support life of the product. The design refresh plan generated by MOCA consists of the number of design refresh activities, their content, and respective calendar dates that minimize the life cycle sustainment cost of a system.

The MOCA model takes as its input the bill of materials (BOM) for a given system, along with the procurement cost and forecasted obsolescence dates or procurement lifetimes of the individual components. MOCA can model multiple levels of hierarchy so that an entire system or system-of-systems containing common components may be loaded into MOCA for concurrent analysis. MOCA also requires a production/deployment schedule as an input. This production/deployment schedule will contain the manufacturing dates and quantity of system instances to be manufactured and deployed. This schedule may be supplemented with the inventory status for components and a forecast of required spares. Using this information, MOCA creates a time line that includes all the projected obsolescence dates for the components, all manufacturing activities, and end-of-support dates (see Figure 9-11).

MOCA supports a design through periods of time when no parts are obsolete, followed by multiple part-specific obsolescence events. When a part becomes obsolete, some type of mitigation approach must take effect immediately. If sufficient inventory exists, a lifetime buy of the part is made or some other short-term mitigation strategy is used that only applies until the next design refresh.

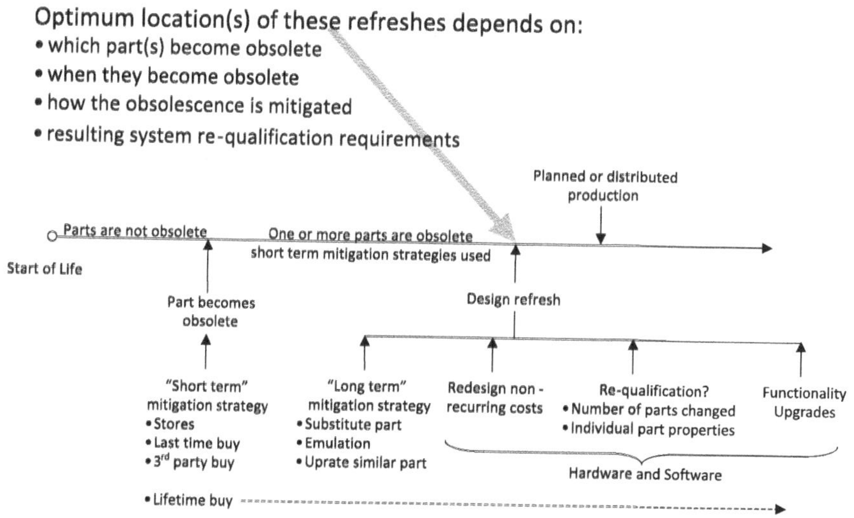

FIGURE 9-11 Design refresh planning analysis time line (presented for one part only for simplicity; however, in reality, there are coupled parallel timelines for many parts, and design refreshes and production events can occur multiple times and in any order).

Next, there are periods of time when one or more parts are obsolete, and short-term mitigation approaches are in place on a part-specific basis. When design refreshes are encountered, the change in the design at the refresh must be determined, and the costs associated with performing the design refresh are computed. At a design refresh, a long-term obsolescence mitigation solution is applied (until the end of the product life or possibly until some future design refresh), and non-recurring, recurring, and requalification costs are computed.

Requalification may be required, depending on the impact of the design change on the application. The necessity for requalification depends on the role that the particular part(s) play and the quantity of noncritical changes made. The last activity appearing on the time line is production. Systems often have to be produced after parts begin to go obsolete because of the length of the initial design/manufacturing process, additional orders for the system, and replenishment of spares.

MOCA is implemented as a discrete event simulator that calculates a system's life cycle cost by traversing the time line described in Figure 9-11 and accumulating system costs as they occur. MOCA generates candidate refresh plans consisting of zero refreshes (all reactive mitigation), exactly one refresh date in the procurement lifetime of the system, exactly two refresh dates, and so on, and places each of the candidate refresh plans onto a time line. The life cycle cost of the system with each of the plans is computed and the candidate plans are ranked, with the lowest life cycle cost being the key performance indicator. Cost model, date, and quantity uncertainties are accommodated in the process via a Monte Carlo simulation (see Singh and Sandborn, 2006, for details).

The output from the model is the minimum life cycle cost refresh plan: a combination of refresh dates and specific refresh content, or the optimum frequency of refreshes. MOCA has been extended to include constraints that limit the time periods when refreshes are allowed to take place (Nelson III and Sandborn, 2012). Refreshes may be constrained by technology road maps, organizational policies, and resource availability.

Value is usually gained from the refresh planning models (and from MOCA specifically) through the identification of cost avoidance opportunities (opportunities to avoid future sustainment costs) associated with optimal planning of refreshes (or the optimal frequency to refresh a system); optimal mixing of reactive DMSMS mitigation solutions with design refreshes; or by identifying refresh points early enough so that appropriate budgets and resources can be put in place. The MOCA methodology can be used during the original product design process or during system sustainment for decision making (for example, when a design refresh is under way, to determine what the best set of changes to make given an existing history of the product and forecasted future obsolescence and future design refreshes); see Singh and Sandborn (2006) for refreshing planning analyses using MOCA.

9.4.9.3 Material Risk Index (MRI) Model The goal of an MRI model is to evaluate the time-dependent risk of a particular function or subsystem within a system being impacted by obsolescence to specific degrees that require specific actions. This evaluated risk can then be mapped to life cycle cost or sustainment dollars at risk.

To perform an MRI on a system (Sandborn, 2012), first, a catalog of functions, subsystems, or specific part profiles is formed. For example, the catalog could contain memory modules, processor boards, and so on. Each profile is characterized by a set of time-dependent obsolescence risk impacts. The periods can represent whatever time frame is relevant to the function or subsystem (usually three or five years). The obsolescence risk (OR) can be interpreted using any one of the following:

- Period-independent model = OR is used to determine the average number of items of a particular profile that are impacted by obsolescence to the extent that some action is required during a period.
- Fractional sum model = OR is the percent of "up-to-date" items that experience obsolescence problems severe enough in the present period to require some action in the present period.
- Probabilistic model = OR is the probability of an item encountering obsolescence problems in the period.

Each of the risk models shown represents a different interpretation (and thereby a different accumulation) of the obsolescence risk values.

Once a catalog has been created, cost models for each type of action that appears in the catalog are developed. Activity-based cost (ABC) models for organizations are the perfect source for data to characterize the costs of activities. Application-specific results are obtained from the MRI model as a function of time using some version of (Sandborn, 2012),

$$\text{Cost in period } i = \sum_{p=1}^{n} (N_p)(OR_{p_i})(C_{p_i}) \qquad (9\text{-}5)$$

where a profile can represent a function, subsystem, or part type, and

n = total number of profiles in the application
N_P = number of instances of the profile p in the application
OR_{Pi} = OR for profile p in period i
C_{Pi} = cost of the action defined in profile p in period i

MRI models require significant resources to create and calibrate, but once created, they are very easy and quick to use.

9.4.10 Open Systems

The rate at which obsolescence is affecting equipment manufacturers is accelerating as the market life of commercially available components decreases. Thus, the burden associated with managing this change is also increasing. To minimize this burden, approaches to design that ease this problem are becoming a necessity. One such approach is open systems. There are many levels at which open systems can be defined. From the equipment manufacturer's point of view, open systems is an approach that permits modules to be replaced as an obsolescence fix, rather than continuing to address the issue at the component level.

Open architecture and standardized interfaces permit form, fit, and function replacement modules to be substituted during repair for those that have become unrepairable due to obsolescence. The existence of standardized interfaces minimizes the effects of changes and sets up change containment boundaries within which change can proceed independently. The common approach to commercial-off-the-shelf (COTS) parts is the use of industry standard modules of a particular family that are purchased from a third party or manufactured directly by the equipment manufacturer and assembled into a chassis. These all share the attribute of adhering to public specifications with strictly defined interface characteristics in terms of mechanical, electrical, and protocol considerations. These modules are both electrically and mechanically "standard," and are available from more than one source, although there may be differences that are not significant to the end application. The modules usually include a number of functions that are not needed for a particular application. They thus carry a cost, size, weight, and power consumption overhead. They also carry verification overhead through the necessity of showing that the unused functions cannot cause unintended effects under reasonably expected operating conditions and scenarios. This approach to open systems is popular, but the previously mentioned drawbacks need to be fully considered when choosing which open systems approach to adopt.

The open systems approach is being actively pursued by military avionics manufacturers (Walden et al., 2000) and is a draft policy of the U.S. DoD (U.S. DoD, 2000; 2002a). Certainly, open systems promises to radically change the business model of equipment manufacturers and rearrange the industry into distinct hardware, software, and systems integrators.

9.4.11 Hardware-Software Independence

A major concern with hardware changes is the effects on a system through faulty operation of previously correct software. Such faults are very subtle, hard to predict, and may only show up under very constrained circumstances that testing has little hope of revealing. In an avionics context, this is the topic that FAA (1999) addresses. A solution to this problem may be using a layered architecture of hardware, hardware services software, applications software, and well-defined and stable specifications for the interfaces. Such architecture has been promulgated in previous work for avionics (ARINC, 1991). Figure 9-12 shows one such avionics implementation, where APEX is the interface definition for software services and COEX is the interface definition for hardware services.

9.4.12 Responsibilities of Customers and End Users

End users often consider obsolescence problems as problems that should be managed by their equipment's manufacturers. It is expected that equipment

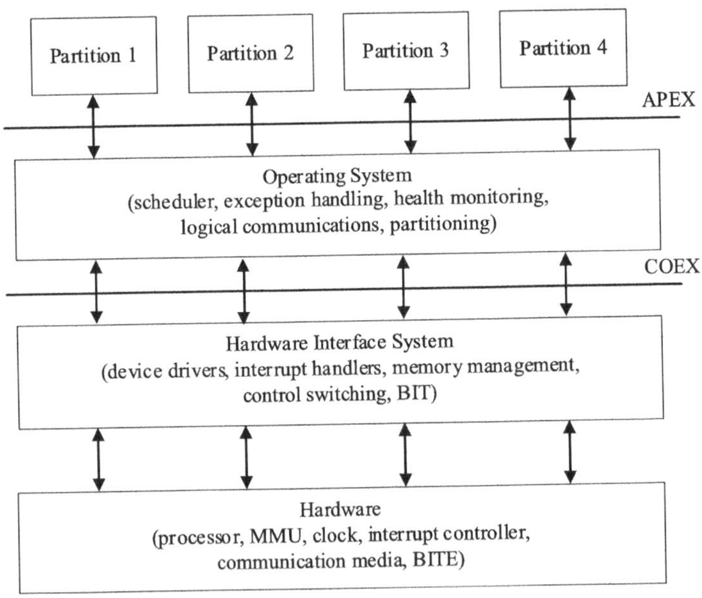

FIGURE 9-12 ARINC 653 APEX software architecture (ARINC, 1997).

manufacturers will continue to provide spare components to support the end user's repair shop operations. The occurrence of an obsolescence event prevents the equipment manufacturer from providing the necessary spare components. The increasingly sophisticated assembly and testing needs of military electronic products are making it difficult and expensive for an end user to provide the necessary supporting repair infrastructure (manufacturing and test equipment and supplies of approved components). The time is rapidly approaching where it will not be economically feasible for end users to perform repairs; end users will have to effectively give up their own service capability in favor of returning failed modules to the equipment manufacturer for repair.

The support commitment provided by equipment manufacturers should be interpreted as a commitment to support the function (not a particular hardware design). This then provides flexibility for equipment manufacturers to manage obsolescence by an open systems approach, provided the equipment manufacturer can develop form, fit, and function replacement items. A continuous program of product update is likely to be the only long-term solution available to the industry.

End users will need to be receptive to proposals to modify the operating envelope of the product to permit economical obsolescence solutions to be implemented. For example, where the best solution would be the use of a component with a different, lower operating temperature limit, it may be possible to extend the start-up time specification to allow the product to reach an in-limit temperature. Such solutions will have to be considered on a case-by-case basis.

Current solutions are often not cost effective. Traditional product support techniques, operating procedures, and maintenance practices may not be compatible with future component trends, and all portions of the supply chain must cooperate. Some solutions will inevitably impact the end users, so they should participate in the development of alternatives in order to realize an overall industrywide solution that is cost effective.

9.5 MONITORING AND CONTROLLING STAGE

The monitoring and controlling stage is usually called the competition stage and describes problem solving and final control of a project (Olfert, 2008). Applied to comprehensive strategic obsolescence management, this stage is used to define, analyze, and evaluate the costs of obsolescence management, as discussed in the following sections.

9.5.1 Economics of Obsolescence Management Strategies

A decision on the choice of an obsolescence management option should be made based on the life cycle cost impact of initiating and implementing the decision. Information about the immediate piece-part and non-recurring engineering costs is not enough. This section describes some of the factors that play a role in determining the life cycle cost of obsolescence management.

The various affected parties view the costs of part obsolescence differently, and the metrics used to measure the costs vary according to the different interests of the entities involved in the process: the suppliers, customers, and acquisition and regulatory agencies. The cost of managing obsolescence also varies depending on the action taken, from relatively low to very high (see Chapter 7). The process of handling obsolescence can even result in long-term cost savings. The task of the parts selection and management team is to choose solutions that result in the optimal life cycle system cost. The various components to be considered when assessing the cost of obsolescence are introduced briefly next.

All solutions to obsolescence involve the cost of managing the system in the future, in addition to the cost of the solution itself. The solutions may manifest themselves in ways that increase the cost of ownership, such as:

- Decreased reparability
- Decreased availability
- Reduced mission effectiveness
- Configuration management
- Support equipment proliferation
- Lost installation baseline controls

The process of responding to obsolescence has specific costs associated with it, such as:

- Redesign or emulation cost
- Time to procure replacement parts
- Additional cost or premium paid for replacement
- Testing costs (for part, board, box, system, and platform)
- Qualification costs (for part, board, box, system, and platform)
- Manufacturing and repair labor costs
- Cost of money

Not all of these costs will be incurred under all situations. Estimating the costs of obsolescence is the process of comparing alternatives in time and technology. Some of these cost alternatives and assessments are discussed in previous sections, in which different strategies for obsolescence management are presented. For each alternative, the cost of various options is to be compared against all other options. When there are a large number of available options, the decision-making process can be very complicated. The importance and difficulty of this process of cost estimation and optimization reemphasizes the importance of the parts selection and management team and process. Determination of the most cost-effective, post-obsolescence solution, typically a solution involving refresh of the product design versus a last-time (bridge) buy, involves both technical and economic tradeoffs: technical tradeoffs to understand the options available to resolve the obsolescence issue and economic tradeoffs to understand the cost-benefit relationships and the most cost-effective time to initiate a redesign activity. For example, Boeing Electronic Products (Porter, 1998; Sandborn, 2012) has developed for internal use a simple set of relationships to allow the product design engineering team to assess the cost-effectiveness of the solution and determine the optimum year to redesign for a minimum cost solution (see Section 9.4.9.1).

Variations in obsolescence management costs come from three major sources: part-level sources, system-level sources, and non-system-specific economic factors. These three sources are briefly discussed in the next section.

9.5.2 Cost Variations at the Part or Component Level

In response to obsolescence, part-level changes are often made by different means, including piece-part replacement from alternative sources and redesign of the system to use a different part. The impact of the choice made at this level goes beyond the direct cost of replacement or cost of man-hours for doing redesign.

For electronic components, for example, a given substitute part for a given board may not have equivalent functionality in all locations on the board. Some locations might require local adjustments, such as adjusting digital timing in asynchronous design applications. Such adjustments add to the cost of the substitution process.

However, careful planning may reduce the overall cost of such changeovers. The design and procurement cost of similar parts can be shared across the platform. The redesign of one board might release a long enough inventory of parts for life cycle maintenance of another board. Board-level redesign can reduce the number of part types that must be maintained in inventory, reducing inventory costs.

9.5.3 Cost Variations at the System or Module Level

Some decisions regarding cost estimation for obsolescence management can only be taken at the system level or higher. Examples of such cost variations are given next.

Cost estimates should also factor in the level of coverage required in the test vectors for validating the various levels of assemblies when making any type of modification. The options are either to aim for very high fault coverage at each level or to merely have adequate coverage to guarantee performance at the next higher level of assembly. Another set of options has to do with deciding at what level form, fit, and function will be maintained. At each level of design, the changes may be made transparent to the next level of assembly and connectors, or some changes can be carried over to the level above.

9.5.4 Cost Variations due to Economic Policy Factors

Contributors to the difficulty of accurately estimating the costs of obsolescence management are macroeconomic factors and public and business policies. The cost of supporting a system (with or without modifications) changes with time. This support cost depends on economic factors such as rate of inflation, cost of capital, and interest rate. The level of available funding for different budget segments also influences the decisions made regarding how to handle obsolescence.

9.5.5 Cost-Benefit Analysis

Once a discontinued item is identified, or the potential obsolescence of a high-risk item or part needs to be avoided, the manager carries out a four-step process to learn the extent of the problem and determine a resolution.

- The first step is to determine where the discontinued item is used and determine the total future requirements for the part in each application.
- The second step is to determine whether a quick fix, such as sufficient stock on hand, substitute parts to cover future requirements, or a bridge buy, is feasible.
- If a quick fix is not feasible, the third step is to gather logistical information about the part. This includes demand levels, future requirements, prequalified sources, and the availability of descriptive information such as specifications, performance data, and interface descriptions.
- Finally, the manager decides which solution or combination of solutions will solve the obsolescence problem and then implements the solution.

A complete analysis of each option is costly and time consuming. An analysis of each option must consider factors such as cost and execution time frames. Such factors, together with the criteria for choosing the optimum obsolescence resolution, are listed at the end of Chapter 7.

Often it is necessary to prove the advantages or disadvantages of different solutions or resolutions with the help of specific management or financial ratios. A value ratio that shows the inefficiency or efficiency of obsolescence management on an economic basis is often used. Therefore, the cost-benefit analysis, in which impacts are expressed in monetary terms, is used with its formulas, as shown in the Equations 9-6 and 9-7 (Thommen and Achleitner, 2005; Holzner, 2006).

$$\text{Benefit Cost Ratio} = \frac{\text{Present Value of Benefits}}{\text{Present Value of Costs}} \tag{9-6}$$

$$\text{Net Benefit} = \text{Present Value of Benefits} - \text{Present Value of Costs} \tag{9-7}$$

If the quotient of the benefit cost ratio shows a result greater than 1, this indicates the efficiency of an obsolescence resolution; in contrast, a benefit cost ratio less than 1 indicates inefficiency (Thommen and Achleitner, 2005; Holzner, 2006). Each obsolescence management option should be considered in relation to other options of solving or avoiding obsolescence issues to determine the best option with the highest benefit cost ratio.

After the best solution is discovered, the net benefit shows how high the specific cost benefit of a specific solution or resolution is. The higher the net benefit, the better the solution.

9.6 STRATEGIC OBSOLESCENCE MANAGEMENT GUIDELINES

Figure 9-13 provides the strategic obsolescence management guidelines regarding how to implement a strategic obsolescence management system within an organization affected by obsolete parts.

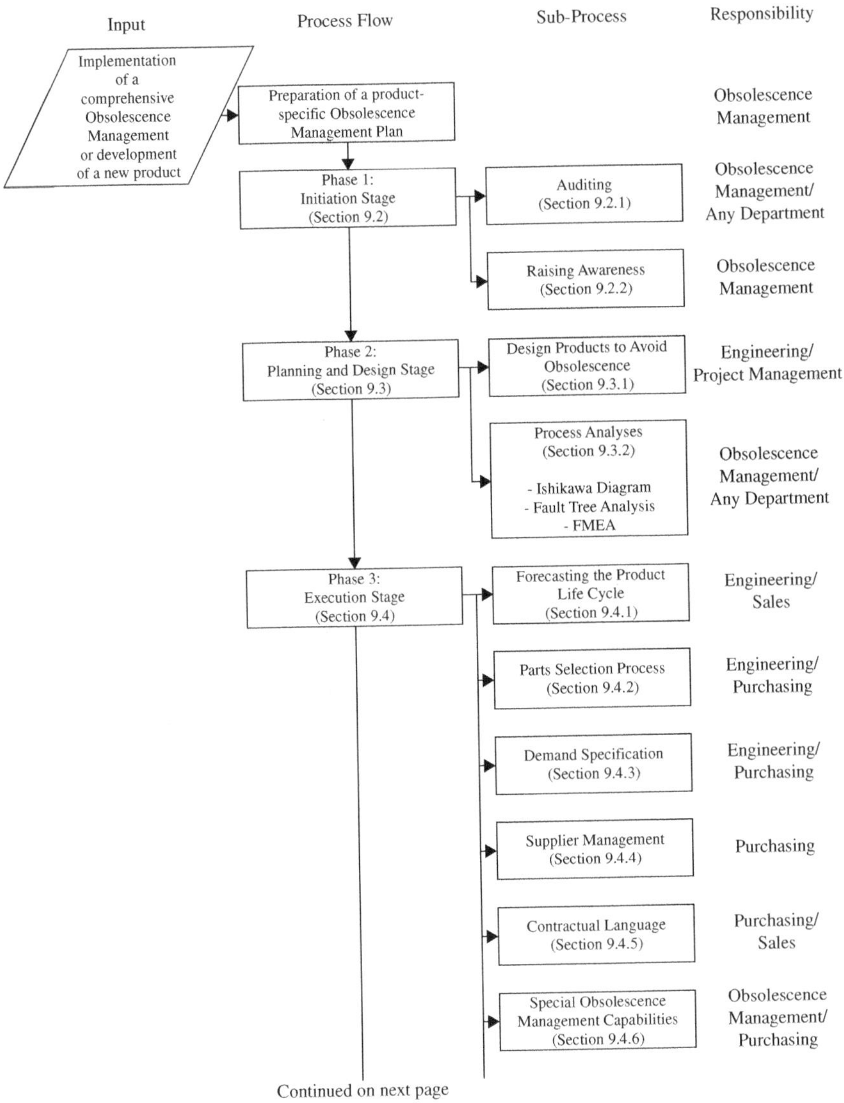

FIGURE 9-13 Strategic obsolescence management guidelines.

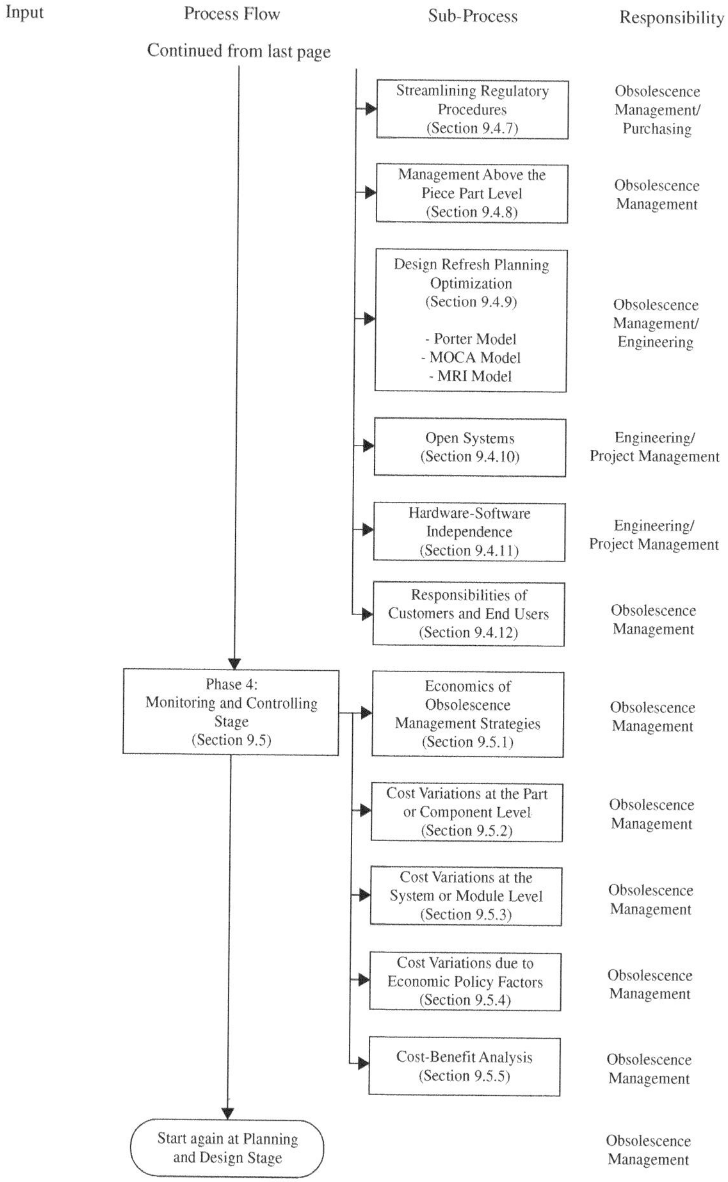

FIGURE 9-13 Continued

Chapter **10**

Obsolescence Management Standards and Organizations

There are standards and guidance documents available for obsolescence management. Several of these are described in this chapter. Since there are no "cookbook" approaches that are viable, the material that is available tends to be of an advisory or guidance nature, even though the title of this chapter uses the word "standard." Obsolescence is an all-pervasive factor in modern system design and support, and it requires managers to make informed tradeoff and optimization decisions. Often there is no single right answer. Organizations that are successful in managing obsolescence are those that consider the impact of obsolescence at all stages of the product life cycle (PLC). Existing standards can do no more than provide guidance to component users. Ultimately, obsolescence is driven by the commercial desires of the component manufacturers, for whom obsolescence management guidelines and standards are not necessarily relevant.

10.1 HELPFUL STANDARDS FOR OBSOLESCENCE MANAGEMENT

The sections that follow describe some of the best known resources on obsolescence and diminishing manufacturing sources and material shortages (DMSMS) management.

Strategies to the Prediction, Mitigation and Management of Product Obsolescence, First Edition. Bjoern Bartels, Ulrich Ermel, Michael Pecht and Peter Sandborn.

10.1.1 Defense Standardization Program Office (SD-22)

In September 2010, the Defense Standardization Program Office (DSPO), published the new SD-22, "Diminishing Manufacturing Sources and Material Shortages (DMSMS): A Guidebook of Best Practices and Tools for Implementing a DMSMS Management Program."

The SD-22 guidebook presents the management guidelines that the majority of the DMSMS management organizations in the world follow. It is a compilation of the best reactive and proactive practices from across the DoD for managing the risk of obsolescence. It also identifies measurement tools that may be useful for analyzing and tracking the effectiveness of DMSMS programs. The guidebook addresses obsolescence issues for both electrical and mechanical parts. Program managers should use this guidebook to quickly pinpoint key actions required in managing DMSMS issues and concerns. Additional information can also be found at the DMSMS Knowledge Sharing Portal (www.dmsms.org) (SD-22, 2010).

10.1.2 Electronic Industries Alliance (EIA)

The Electronic Industries Alliance (EIA) has published standard EIA-724 (EIA-724, 1997) to help organizations define product life cycles. This document provides a system of product life cycle coding (PLC) that separates a product's life cycle into the six categories defined in Chapter 3.

The Government Electronics and Information Technology Association (GEIA, a division of the Electronic Industries Alliance) G-12 Solid State Devices Committee has developed a set of obsolescence management practices that can be used by original equipment manufacturers (OEMs) during the design and development of electronic systems to mitigate the effects of DMSMS. GEB1, a GEIA technical paper presenting an overview of obsolescence management (EIA, 2002), includes proactive DMSMS mitigation methods, such as technology independence (use of VHDL, software portability), technology road mapping, technology insertion, planned system upgrades, life cycle analysis, and DMSMS monitoring. GEB1 also addresses traditional responses to obsolescence events, such as alternate sourcing, substitution, redesign or design modification, reverse engineering, and reclamation. GEB1, Diminishing Manufacturing Sources and Material Shortages (DMSMS) Management Practices, is available through Global Engineering Documents.

Further, in 2005 GEIA published the standard "STD-0002-1—Aerospace Qualified Electronic Component (AQEC) Requirements, Volume 1—Integrated Circuits and Semiconductors." This standard applies to integrated circuits and semiconductors that exhibit a minimum set of requirements or to information provided by the part manufacturer that will allow a standard COTS component to be designated as AQEC by the manufacturer (STD-0002-1, 2005).

EIA also hosted the activities of the Avionics Process Management Committee (APMC), which further developed IEC documents on component management and uprating (EIA-4899, 2002; EIA-4900, 2002).

10.1.3 Joint Electron Device Engineering Council (JEDEC)

The Joint Electron Device Engineering Council is a standards body set up by companies that manufacture semiconductors. They are affiliated with the Electronic Industries Alliance (EIA). JEDEC has published the following relevant standards:

- JESD53, "Management of Component Obsolescence by Government Contractors," describes the elements of a component obsolescence management program (JESD53, 1996). This is a very high-level document and does not go into much detail about how a plan should be developed.
- JESD48, "Product Discontinuance," sets out the minimum standards that a semiconductor manufacturer should adhere to when issuing obsolescence notices (JESD48, 1997). This document covers such items as content, timing, last-time buy (LTB) options, and aftermarket sources, and establishes guidelines for customer notification of the discontinuance of semiconductor parts. The goal of this standard is to aid customers in managing end-of-life (EOL) supply issues and transitioning to replacement parts. Suppliers are requested to provide a minimum of 6 months advance notice for discontinuance of multiple-source parts and 12 months for single-source parts. This notice should be provided in writing to all affected direct customers who have purchased the parts being discontinued during the past two years, all direct sales channels, and all authorized distributors. Distributors then notify their own customers of the planned part discontinuance as per their own policies. At a minimum, the notification should include the last day for which purchase orders will be accepted and the affected part numbers. Upon request, the manufacturer should also provide information and technical data to help customers find replacement sources for discontinued parts.

10.1.4 International Electrotechnical Commission (IEC)

The International Electrotechnical Commission (IEC) is made up of experts in electricity and electronics who participate in technical committees providing global standards to the world's electrotechnical industries (IEC, 2010).

In 2007, the IEC published the international standard "IEC 62402 Ed. 1.0: Guide to Managing Obsolescence," which provides guidance for establishing a framework for obsolescence management and for planning a cost-effective obsolescence management process that is applicable through all phases of the product life cycle (IEC-62402, 2010).

Although not strictly an obsolescence standard, the IEC has published a guidance document on the uprating of electronic parts (IEC-62240, 2001). Uprating is one of the potential solutions to obsolescence (see Chapter 7).

Further, the IEC has published a guide to electronic component management (IEC-62239, 2001) that is also not an obsolescence standard, but nonetheless provides a potential tool for obsolescence management.

Also relevant is "IEC 62435: Electronic Components—Long-duration Storage of Electronic Components," which provides guidelines for long-term storage of obsolete parts. All handling, protection, and storage of parts and test operations are discussed (IEC-62435, 2005).

10.1.5 DIN Deutsches Institut für Normung e.V.

DIN, the German Institute for Standardization, is based in Berlin and offers stakeholders a platform for the development of standards to meet market requirements. DIN is the acknowledged national standards body that represents German interests in European and international standards organizations (DIN, 2010a).

As 90 percent of the standards developed by DIN are international in nature, the DIN, like the IEC, published the standard "EN 62402:2007—Anleitung zum Obsoleszenzmanagement—Obsolescence management (IEC 62402:2007)" in a German version.

10.1.6 British Standards Institution (BSI)

The British Standards Institution (BSI) is the national standards body of the United Kingdom, and it was also the world's first standards body. BSI works with manufacturing and service industries, businesses, governments, and consumers to facilitate the production of British, European, and international standards on a non-profit basis (BSI, 2010b).

The British Standards Society issued a standard called "Obsolescence management—Guide to the substitution of components in electronic equipment—PD 6614:2002," which was first published in June 1999; a second edition was published in June 2002.

10.1.7 STACK International

STACK International is a U.K.-based common interest group of component users. STACK provides a publicly available part procurement specification, "STACK-0001" (STACK, 1999). This specification provides a guideline for product discontinuance notices. This requirement specifies a 6-month end-of-life (EOL) notification of last-time buy (LTB) for multi-sourced parts and 12 months for single-sourced parts.

10.1.8 Electronics Industry Quality Conference (EIQC)

The Electronics Industry Quality Conference has produced a brief guide describing the data that a part manufacturer should provide within obsolescence and last-time buy (LTB) notices. This document requires parts suppliers to give at least one year's notice of discontinuation and provide dates for the last production run, unlimited quantity last-time buy date, last delivery date, and any suggested alternate parts (EIQC-0002, 1993).

10.1.9 Airlines Electronic Engineering Committee (AEEC)

The Airlines Electronic Engineering Committee (AEEC), operated by Aeronautical Radio Inc. (ARINC), is a standards-setting body created by the world's airlines. It primarily publishes performance standards for avionics products. AEEC completed its report number 662, "Strategies to Address Electronic Component Obsolescence in Commercial Aircraft" (ARINC, 1999a), in December 1999 and issued it as an approved document. This document establishes guidelines that should be observed during the initial design, production, and maintenance of avionics line replaceable units (LRUs) and presents short- and long-term strategies to minimize the costs associated with the obsolescence of electronic components used in fielded LRU designs.

10.1.10 VMEbus International Trade Association (VITA)

Commercial database providers generally only have data for electronic piece parts. Although many commercial databases claim to address electronic assemblies, the reality is that there is relatively little or no commercial forecasting available for items above the piece-part level. In addition, not every piece part will necessarily be recognized by the commercial database tools. Therefore, obsolescence status and forecasting must be supplemented by researching the parts.

The most straightforward way to gather forecasting data is to consult manufacturers and official distributors directly. To standardize this inquiry, the VMEbus International Trade Association released the open VITA 53.0 Standard for Commercial Technology Market Surveillance in 2009. This standard was chaired by NSWC Crane (U.S. Navy) and defined by a group of experts from various companies (Boeing, General Electric, Foxconn, Bustronic, and others) (VITA 53.0, 2009). This open standard contains the following obligatory questions for the manufacturer's inquiry form and can be individually adapted to the company's specific requirements (see Table 10-1).

10.2 HELPFUL ORGANIZATIONS FOR OBSOLESCENCE MANAGEMENT

Private firms, organizations, official state-owned departments, and universities can provide help and database services for obsolescence management. Some of these are described in the following sections.

10.2.1 U.S. Department of Defense (DoD)

The U.S. DoD and other defense procurement agencies worldwide have a strong interest in parts obsolescence. The response of DoD has been to actively

TABLE 10-1 Topics for Part Researching According to the Open VITA 53.0 Standard for Commercial Technology Market Surveillance (VITA 53.0, 2009)

Basic Facts for Manufacturer Inquiries
1. Product Identification
1.1. Company name, product name, and product part number
2. Product Description
2.1. Please provide a brief description of the product.
2.2. What operating systems are supported?
2.3. If this product is part of a family/series of products that uses a base product, what portion of the part number (if any) identifies the family/series?
2.4. Please provide a brochure, data sheet, or operation manual of this product.
2.5. If firmware is associated with the product, please describe the firmware and give firmware version numbers.
2.6. Is the product Restriction of Hazardous Substances (RoHS) compliant?
2.7. Is the product a matched set to any other hardware?
3. Start of Production
3.1. Please provide the start of production date or the date of market introduction.
4. End of Production
4.1. Please provide an actual or estimated end of production date.
4.2. If the product is no longer being produced, when was production discontinued?
4.3. If the product is no longer being produced, how much stock is available for purchase?
4.4. Has an end-of-life letter (or similar notice) been issued? (If so, please provide a copy.)
4.5. What is the last date this product can be purchased?
4.6. If product is still in production, for planning purposes only, what is the planned or estimated end of production date?
5. End of Support
5.1. Please provide an actual or estimated end of support date.
5.2. How long after end of production will (the interviewer) be able to have this product repaired?
5.3. What is the condemnation (scrap) rate of the product?
6. Product Cost
6.1. What is the cost of the product? List pricing options and volume discounts.
6.2. What is the average repair cost?
7. Failure Rate
7.1. What is the mean time between failure and how is it calculated? (Hours and method: Please state whether predicted or measured.)
8. Recommended Replacement
8.1. Is there a recommended replacement for this product?
8.2. If yes, please provide part number and list price.
8.3. If yes, how difficult (in your opinion) will it be to replace the old part with the new part (direct, minor, or complex)? If it is known that this varies depending on the application, please describe the variances.

TABLE 10-1 Continued

Basic Facts for Manufacturer Inquiries
9. Warranty
9.1. Describe the provisions for the standard warranty for this product.
9.2. Is an extended warranty offered? If so, what are the cost, length, and conditions?
9.3. Do you enter into maintenance agreements? If so, can you provide general costing information?
9.4. What trade-in value options do you offer (if any)?
10. Aftermarket Vendor
10.1. Do you plan to enter into an agreement with an aftermarket vendor when you discontinue production or support of this product?
10.2. If yes, what is the manufacturer's name and contact details? (If there are multiple agreements, please list them all.)
10.3. If no, would you be willing to enter into such an agreement?
11. Other Potential DMSMS Case Solutions
11.1. Is this product identified by any other name or part number? If yes, please provide p/n or name.
11.2. Does your company use distributors? If yes, please list all.
11.3. Do you know of other Department of Defense or government programs using this product? If yes, please list all.
12. Firmware Upgrades and Engineering Design Changes
12.1. Does the product contain firmware? If yes, please provide the current revisions of all firmware used on the product (date of change, description/reason for change, revision or version number).
12.2. Does your company offer a notification service for firmware upgrades and design changes for this product? If yes, please describe this service, including the cost of the service (if any), as well as how to subscribe to this service.
12.3. Does your company offer the opportunity to freeze the firmware version on this product? If yes, please provide the cost of this service (if any).
12.4. Is any of the firmware on this product upgradeable by the end user (customer)? If yes to any of the firmware on the product, please describe the process for the end user upgrading the firmware for this product.

promote a transition to commercial items by instituting a sweeping acquisition reform program. In the U.S., the DoD's guidance on obsolescence management for military program managers is contained within the U.S. DoD Defense Acquisition Deskbook (U.S. DoD, 2002b). In addition, the U.S. DoD is sponsoring a number of research projects. For example, the U.S DoD sponsored projects through the U.S. Air Force MANTECH program aimed at providing either guidance documents or software tools (U.S. DoD, 2001a; Bumbalough, 1999).

Military electronics manufacturers supplying parts to specifications MIL-PRF-38535 (MIL-PRF-38535, 1989), MIL-PRF-38534D (MIL-PRF-38534D, 1999), and MIL-PRF-19500 (MIL-PRF-19500M, 1999) have required stricter change-notification procedures. They are based on the Qualified Manufacturers

List (QML) and Government Industry Data Exchange Program (GIDEP). QML dictates change-notification and test requirements, and GIDEP serves as a recommended path for change-notification.

QML establishes the general performance requirements for semiconductor parts and the quality and reliability requirements that must be met. The manufacturer notifies the qualifying activity prior to the implementation of any change in a product or quality assurance program that may affect performance, quality, reliability, or interchangeability. The manufacturer submits engineering data, quality conformance inspection data, or a suggested test plan. The manufacturer also explains the qualification process that ensures that reliability and quality will not be affected and that the product will continue to meet the specification requirements.

10.2.2 Government Industry Data Exchange Program (GIDEP)

The U.S. DoD supports the GIDEP organization, which acts as a clearinghouse for obsolescence information. GIDEP is a program run by the U.S. government that aids in change and obsolescence notification. Participants in GIDEP are provided electronic access to engineering data, failure experience data, metrology data, part information data, reliability and maintainability data, and PCNs. To aid in obsolescence management, DMSMS notices of part discontinuances and information on alternate sources and aftermarket suppliers are provided.

GIDEP is a government-wide central system for sharing technically valid, fact-based information between the U.S. and Canadian governments, as well as supporting industries. GIDEP helps manage part discontinuances by providing a platform for sharing solutions submitted by the GIDEP community and made available through the database. To help locate discontinued parts, GIDEP provides a service called Urgent Data Requests to allow participants to query the 1600-member GIDEP community to identify sources of surplus parts. Only U.S. or Canadian government agencies or supporting industries may join GIDEP. There is no fee to belong to the program and share information. The only cost is to agree to share information and provide GIDEP with an annual utilization report outlining the benefits of belonging to GIDEP.

10.2.3 Defense Logistics Agency (DLA)

The U.S. Defense Logistics Agency (DLA) was founded in 1961 and is an agency in the U.S. DoD, with civilian and military personnel throughout the world. The DLA is an integral part of the United States' military defense. The DLA runs the DMSMS working groups for the U.S. DoD and originated the website of the DMSMS Knowledge Sharing Portal (www.dmsms.org).

10.2.4 Defense Microelectronics Activity (DMEA)

Through its Industry-Government Partnership Initiative, the Defense Microelectronics Activity (DMEA) has developed strategic partnerships with industry and key government organizations in an effort to ensure that the U.S. DoD's critical requirements for microelectronics technologies are met. These relationships serve to bring the DMEA team of microelectronics specialists together with the strength and expertise of the industrial sector to solve complex customer problems. The mission of DMEA is to leverage advanced technologies to extend the life of weapons systems by improving their reliability and maintainability and address the problem of diminishing manufacturing sources (DMS). DMEA has been designated the U.S. DoD's executive agent for integrated circuit (IC) microelectronics obsolescence. DMEA is a key player in the development and coordination of solutions to U.S. DoD's obsolescence problems and is responsible for issues relating to IC microelectronics obsolescence. These responsibilities are across the full spectrum of advanced microcircuit technology and obsolescence issues. They include:

- Providing advice and recommending policy to the Deputy Under Secretary of Defense for Logistics on issues affecting U.S. DoD weapons systems due to IC obsolescence.
- Developing partnerships with the military services, other U.S. DoD and non-DoD organizations, the semiconductor industry, and the electronics industry.
- Fostering cooperation with the manufacturing industry to achieve a joint resolution to current or potential DMSMS issues that affect the U.S. DoD.
- Hosting workshops and conferences to address issues related to IC obsolescence.
- Advocating and developing cost-effective technical solutions to the U.S. DoD's IC microelectronics obsolescence problems.
- Implementing an "IC clearinghouse" for IC problems and solutions to provide a vehicle to talk about common obsolescence problems and share in the solutions.
- Assessing the current state and trends of microelectronics technology and the IC manufacturing industry to develop road maps for satisfying the U.S. DoD's requirements for microcircuit part technology and mitigating the effects of obsolescence.
- Developing guidelines and strategies in conjunction with the military services and the Defense Logistics Agency (DLA), to help weapons system program managers effectively manage and mitigate obsolescence.

10.2.5 UK Ministry of Defence (UK MoD)

The UK Defence Evaluation and Research Agency (DERA) jointly maintains a website with industry for information on obsolescence called the National Obsolescence Centre, located in Bromley, Kent, UK. The UK MoD has also published "DEF-STAN-00-71" (DEF-STAN-00-71, 1998). Although called a standard, this is really a guidance document on obsolescence management.

Feasibility studies are promoted and supported by the UK MoD. For example, one study undertaken in the UK was completed in December 2002. It indicated that UK MoD, UK industry, and other UK obsolescence focal points were willing to participate in the development of a UK version of the cost metrics by providing resolution cost data (QinetiQ and ARINC, 2004).

10.2.6 Component Obsolescence Group (COG)

The Component Obsolescence Group (COG) (COG, 2010) is a special interest group of like-minded professionals from all levels of the supply chain and across all industries and relevant government agencies concerned with addressing and mitigating the effects of obsolescence. COG promotes a proactive approach to the management of obsolescence and the development of processes for dealing with it through the mutual, voluntary sharing of obsolescence issues, strategies, and solutions.

COG operates as a self-financed body to:

- Discuss the criteria that affect product obsolescence
- Identify and analyze various approaches for addressing obsolescence
- Promote the advantages of proactive management of obsolescence
- Arrange training courses on subjects in the field of obsolescence management
- Organize conferences and trade exhibitions
- Encourage additional parties to join COG
- Liaise and cooperate with other national and international bodies
- Enable data sharing and provide links to available information

A series of meetings are held each year to allow discussion of topics concerning component obsolescence and hear presentations on possible obsolescence solutions.

Websites (www.cog.org.uk and www.cog-d.de) are available that have links to obsolescence information from around the world. Company name changes are tracked by the "Who are they now?" feature. Some of the websites are available to members only. Regular conferences, exhibitions, and training sessions are also arranged.

Furthermore, the COG sells several publicly available booklets on potential minefields in the area of obsolescence in their web-based shop.

The COG was formed in the UK in 1997 and provides a forum for industry professionals to meet each other and exchange information. It also enables data sharing and provides links to available information using modern retrieval methods. Since 2004, COG has also been active in Germany and other European countries.

10.2.7 University of Maryland—CALCE

Some of the research presented in this book was performed at the Center for Advanced Life Cycle Engineering (CALCE) Electronic Products and Systems Center (EPSC) at the University of Maryland, College Park. CALCE is recognized as a founder and driving force behind the development and implementation of physics-of-failure (PoF) approaches to reliability, as well as a world leader in accelerated testing, electronic parts selection and management, and supply-chain management. More than 100 electronic products and systems companies from all sectors, including avionics, automotive, computer, semiconductor, and electronics manufacturers, support CALCE. The Center performs research in the thrust areas of interest to its sponsors and provides information, services, and tools (for example, CALCE Mitigation of Obsolescence Cost Analysis [MOCA]) to help its clients develop competitive electronic products. CALCE researchers have authored more than 35 internationally acclaimed textbooks and well over 1000 research publications relevant to electronics reliability and electronic part supply-chain management, including obsolescence forecasting and management (CALCE, 2010).

10.2.8 Federal Aviation Administration (FAA)

The FAA in the United States manages obsolete parts for many different US government programs. The FAA has issued a document detailing researching COTS assemblies and COTS software (Shaffer and McPherson, 2010).

References

ABI Forecasts Wireless to Grow by Billions, *Advanced Packaging*, vol. *9*, no. 3. March 1, 2000.

Adamson, S. J., CSP and Flip Chip Packaging, *Advanced Packaging*, vol. *9*, January 1, 2000.

Adbusters, Consumer Society Is Made to Break, https://www.adbusters.org/category/tags/obsolescence, accessed on April 5, 2010.

Air Force Research Laboratory, Wright Patterson Air Force Base, Ohio, Research and Development (R&D) Supportability Design Guide, AFRL, 1998.

Ahuja, G., and Katila, R., Technological Acquisitions and the Innovation Performance of the Acquiring Firms: A Longitudinal Study, *Strategic Management Journal*, vol. *22*, pp. 197–220, 2001.

Allocation Components BlogSpot, Electronic Components: Allocation, Lead Times & Shortages, http://allocationcomponents.blogspot.com/, accessed on April 11, 2010.

Allocation Components BlogSpot, Tragic Japan Earthquake Affects Some Electronic Semiconductor Manufacturing Facilities, http://allocationcomponents.blogspot.com/2011/03/tragic-japan-earthquake-affects-some.html, accessed on July 6, 2011.

Apple Inc., Macbook Air Product Website, http://www.apple.com/de/macbookair/specs.html, accessed on October 23, 2010.

APT News, The Bill Gates Method, July 21, 2003, http://www.placementpartner.com/apt%20news/news7-21-03.html.

Angel Investor News, Thought Provoking Advice, http://www.angel-investor-news.com/Articles-Advice-To-Ponder.htm, accessed on April 6, 2010.

Strategies to the Prediction, Mitigation and Management of Product Obsolescence, First Edition. Bjoern Bartels, Ulrich Ermel, Michael Pecht and Peter Sandborn.

ARINC Report, *Design Guide for Integrated Avionics*, ARINC Report 651, Aeronautical Radio Inc., Annapolis, MD, November 1, 1991.

ARINC Report, *Avionics Application Software Standard Interface*, ARINC Specification 653, Aeronautical Radio Inc., Annapolis, MD, January 1, 1997.

ARINC Report, *Strategies to Address Electronic Component Obsolescence in Commercial Aircraft*, ARINC Report 662, Aeronautical Radio Inc., Annapolis, MD, December 1, 1999a.

ARINC Report, Aeronautical Radio Inc. Annapolis MD, Resolution Cost Factors for Diminishing Manufacturing Sources and Material Shortages, February 1, 1999b, http://www.dmea.osd.mil/dmsms_resolution_cost_factors-all.pdf, accessed on January 15, 2002.

ARM Ltd, Product Website on Embedded Memory SRAM IP, http://www.arm.com/products/physical-ip/embedded-memory-ip/sram.php, accessed on November 2, 2010.

Arnold, H., Elektroniknet.de, Programmierbarkeit und Innovationsfreude, http://www.elektroniknet.de/bauelemente/news/article/31159/0/Programmierbarkeit_und_Innovationsfreude/, accessed on December 22, 2010.

Atmel Cooperation, Product Website of AT25DF161, http://www.atmel.com:80/dyn/products/product_card.asp?part_id=1890, accessed on November 16, 2010a.

Atmel Cooperation, Product Obsolescence Information Website, http://www.atmel.com/dyn/quality/quality_prod_obsl_result.asp, accessed on November 22, 2010b.

Atmel Cooperation, Product Website and Datasheet of AT45DB161D, http://www.atmel.com/dyn/resources/prod_documents/doc3500.pdf, accessed on November 22, 2010c.

Atterbury, R., Component Obsolescence Group, *United Kingdom; Minenfeld Obsolescence, A Guide to Tackling Disappearing Products*, Issue 2, 2005.

Avnet Abacus Lead Time Prognosis Overview, Monday, March 15, 2010, http://www.avnet-abacus.eu/services/lead-time-overview/, accessed on April, 11, 2010.

Baca, M., QinetiQ Technology Extension Corporation, April 2010.

Baer, D., Wermke, M., Duden - das grosse Fremdwoerterbuch. Herkunft und Bedeutung der Fremdwoerter. mit "umgekehrtem" Woerterbuch: deutsches Wort - Fremdwort. 2., neu bearb. und erw. Aufl. Mannheim: Dudenverlag, 2000.

Bartsch, H.-J., Taschenbuch Mathematischer Formeln, S.599, 16. Auflage, Fachbuchverlag Leipzig, 1994.

Battersbury, C., Component Obsolescence Group, United Kingdom, The Date Coding Minefield. A Guide to Component Traceability through the Date Code, 2003.

Battersbury, C., Component Obsolescence Group, United Kingdom, The Counterfeit Electronic Components Minefield, A Guide to Understanding and Avoiding Counterfeit Components, Issue 1, 2008.

Bayus, B., An Analysis of Product Lifetimes in a Technologically Dynamic Industry. *Management Science*, vol. *44*, pp. 763–775, 1998.

Berka, S., Operating Systems Website, http://www.operating-system.org/betriebssystem/_german/bs-winnt31.htm, accessed on October 12, 2010.

Biagini, R., Rowland, M., Jackson, M., and Pecht, M., Tipping the Scales in Your Favor When Uprating, *Circuits & Devices*, pp. 15–23, July 1999.

Boston Consulting Group, The Product Portfolio, January 1970, http://www.bsg.com/expertise_impact/publications//PublicationsDetails.aspx?id=tcm:12–13257, accessed on April 3, 2010.

Brewin, A., Component Obsolescence Group, United Kingdom, The Pb-free Minefield, A Guide to Mitigating Risk during the Transition to RoHS Compliance, 2005.

Brinton, J. B., SRAM: Integration Taking Starch out of Unit Growth, Semiconductor Business News, 06. January 2000, CMP Media, http://www.siliconstrategies.com/story/OEG20000106S0010, accessed on January 15, 2002.

Bronstein, I.N., Semendjajew, K.A., Taschenbuch der Mathematik, 24. Auflage, page 664.

Brown, G., Lu, J., and Wolfson, R. 1964. Dynamic Modeling of Inventories Subject to Obsolescence. *Management Science*, vol. *11*, no. 1, pp. 51–63, 1964.

Brown, P., IBM Says No to Rambus, *Electronic News*, May 31, 1999, Cahners Business Information, http://www.e-insite.net/electronicnews/index.asp?layout = article& articleid = CA71026&pubdate = 5/31/1999, accessed on January 15, 2002.

Brown, W. D., and Brewer, J. E., *Nonvolatile Semiconductor Memory Technology: A Comprehensive Guide to Understanding and Using NVSM Devices*, New York: IEEE Press, 1998.

BSI British Standards Society, About BSI British Standards, http://www.bsigroup.com/en/Standards-and-Publications/About-BSI-British-Standards/, accessed on November 19, 2010a.

BSI British Standards Society, Obsolescence Management—Guide to the Substitution of Components in Electronic Equipment, PD 6614, June 16, 2002b.

Bumbalough, A., USAF Manufacturing Technology's Initiative on Electronics Parts Obsolescence Management, Proceedings of 44th International SAMPE Symposium, pp. 2044–2051, May 1999.

Cahners Business Information, Process Pedal to the Metal, *Electronic News*, February 8, 1999, http://www.e-insite.net/electronicnews/index.asp?layout = article&articleId = CA71985, accessed on January 14, 2002.

Cahners Business Information, EDN's 27th Annual Microprocessor and Microcontroller Directory, Electronic Design News, 14. September 2000a, Cahners Business Information, http://www.e-insite.net/ednmag/index.asp?layout = article&articleId = CA47177&stt = 001, accessed on January 15, 2002.

Cahners Business Information, DSP Market Growth Is Strong; $4 Billion Barrier Broken, *Electronic News*, 2000b.

CALCE, Center for Advanced Life Cycle Engineering, About CALCE, http://www.calce.umd.edu/general/general.htm, accessed on November 19, 2010.

Carbone, J., Growth Means More Consolidation, More Services, *Purchasing Magazine*, October 19, 2000, PennWell Corporation. http://www.manufacturing.net/pur/index.asp?layout = articleWebzine&articleid = CA139164, accessed on January 21, 2001.

Carbone, J., Buyers Are Putting the Pieces Together, Purchasing Magazine, 22. March 2001. PennWell Corporation, http://www.manufacturing.net/pur/index.asp?layout = articleWebzine&articleid = CA139772, accessed on January 14, 2002.

Carbone, J. Where Are the Parts? *Purchasing*, vol. *132*, pp. 44–47, 2003.

Cattani, K. D., and Souza, G. C., Good Buy? Delaying End-of-Life Purchases, *European J. of Operational Research*, vol. *146*, pp. 216–228, 2003.

Cavill, P. J., Whole Program Life COTS: Warding Off the Curse of Component Obsolescence, Military & Aerospace Electronics, 01. December 2000. PennWell Corporation, http://mae.pennwellnet.com/Articles/Article_Display.cfm?Section = Archives& Subsection = Display&ARTICLE_ID = 88120, accessed on January 15, 2002.

Chakravarti, A. K, Vasanta, B., Krishnan, A.S.A., and Dubash, R. K., Modified Delphi Methodology for Technology Forecasting: Case Study of Electronics and Information Technology, *Technological Forecasting & Social Change*, vol. *58*, no. 1–2, pp. 155–166, 1998.

Christiansen, D., *Electronics Engineer's Handbook*, New York: McGraw Hill, 1997.

CIE, Does Software Go Obsolete?, *Components in Electronics*, March 2006.

CMP Media, Top Semiconductor Suppliers 2000 MCUs, *Electronic Business*, March 1, 1999, http://img.cmpnet.com/ebn/graphics/0601/topsemimcu.pdf, accessed on January 25, 2002.

CMP Media, Toshiba Aims Stacked 8-Mbit SRAM/64-Mbit Flash at Cell Phones, *Semiconductor Business News*, March 9, 2000a, http://www.siliconstrategies.com/story/OEG20000309S0004, accessed on January 15, 2002.

CMP Media, TSMC MoSys Fabricate Ultra-dense Embedded SRAM on 0.15-micron Process, *Semiconductor Business News*, May 17, 2000b, http://www.siliconstrategies.com/story/OEG20000517S0035, accessed on January 15, 2002.

CMP Media, UMC Claims Foundry Technology Lead with Copper SRAM Made with 0.13-Micron Process, *Semiconductor Business News*, May 12, 2000c, http://www.siliconstrategies.com/story/OEG20000512S0017, accessed on January 15, 2002.

CMP Media, DSP Design Trends: TI Retains Leadership Position in DSP Market, *Communication System Design* (Special Section), April 1, 2000d, http://www.csdmag.com/main/2000/04/0004trend2.htm, accessed on January 15, 2002.

CMP Media, Copper Tool Spending Will Grow 50% in 2000 Says Report, *Semiconductor Business News*, February 2, 2000e.

CMP Media, With Wafer Reclamation Growing, Company Renames Itself TTI Silicon, Semiconductor Business News, January 19, 2000f, www.siliconstrategies.com/story/OEG20000119S0036, accessed on January 15, 2002.

CMP Media, Microprocessors, *Electronic Business Online*, June 6, 2001a, http://www.ebnonline.com/story/OEG20010604S0102, accessed on January 18, 2002.

CMP Media, Top Semiconductor Suppliers—Analog ICs, *Electronic Buyers News*, June 6, 2001b, CMP Media, http://www.ebnonline.com/story/OEG20010605S0059, accessed on January 18, 2002.

CMP Media, TSMC to Offer MoSys Embedded SRAM Technology in 0.18-micron Process, *Semiconductor Business News*, January 31, 2000, http://www.siliconstrategies.com/story/OEG20000131S0023, accessed on January 15, 2002.

CNET, ASE Acquires Two Motorola Chip Plants, July 6, 1999, http://news.cnet.com/news/0-1006-200-344451.html, accessed on January 15, 2002.

COG, Component Obsolescence Group, Introductory Presentation, http://www.cog.org.uk/PDF/cog%20presentation%20(2).pdf, accessed on November 19, 2010.

ComputerInfoWeb.com, Computer Electronics, Blu-Ray, http://www.computerinfoweb.com/computer_electronics/Blu_Ray.php, accessed on April 5, 2010.

Condra, L., Combating Electronic Obsolescence by Using Common Processes, 2nd Annual NDIA Systems Engineering & Supportability Conference, Sept. 1999.

Condra, L., Anissipour, A., Mayfield, D., and Pecht, M., Electronic Components Obsolescence, *IEEE Transactions on Components, Packaging and Manufacturing Technology*—Part A, vol. *20*, pp. 368–371, 1997.

Condra, L., and Das, D., Junction Temperature Considerations in Evaluating Electronic Parts for Use Outside Manufacturers-Specified Temperature Ranges, *IEEE Transactions on Components and Packaging Technologies*, vol. *24*, pp. 721–728, Dec. 1, 2001.

Condra, L., Followell, D., Houchens, G., Jenks, J., Koehler, M., and Porter, Z., Minimizing the Effects of Electronic Component Obsolescence, *Aero Magazine*, 10, April 1, 2000, Boeing Company. Seattle, WA, http://www.boeing.com/commercial/aeromagazine/aero_10/elect.pdf, accessed on January 14, 2002.

Condra, L., Hoad, R., Humphrey, D., Brennom, T., Fink, J., Heebink, J., Wilkinson, C., Marlborough, D., Das, D., Pendse, N., and Pecht, M., Terminology for Use of Electronic Parts outside the Manufacturers Specified Temperature, *IEEE Transactions on Components and Packaging Technology*, vol. *22*, pp. 355–356, Sept. 1, 1999.

Condra, L., and Matthews, J., The Growing Problem of Component Obsolescence, *Avionics Magazine*, March 1999.

Crane DMS Technology Center, U.S. Navy, GEM: The Generalized Emulation of Microcircuits, Navsea, http://dtc-dms.crane.navy.mil/html/gem.htm, accessed on January 15, 2002.

Cushing, M. J., Mortin, D. E., Stadterman, T. J., and Malhotra, A., Comparison of Electronics-reliability Assessment Approaches, *IEEE Transactions on Reliability*, vol. *42*, Dec. 1, 1993.

Cypress Semiconductor, Inc., Design Guide QDRII, QDRII+,DDRII and DDRII+ Design Guide, AN4065, Document No. 001–15486 Rev.*A, http://www.qdrconsortium.com/technotes/cypress/an4065_12.pdf, accessed on November 1, 2010.

Das, D., Pendse, N., and Wilkinson, C., Parameter Recharacterization: A Method of Thermal Uprating, *IEEE Transactions on Components and Packaging Technologies*, vol. *24*, pp. 729–737, Dec. 1, 2001.

Dasgupta, A., Magrab, E. B., Anand, D. K., Eisinger, K., McLeish, J. G., Torres, M. A., Lall, P., and Dishongh, T. J., Perspectives to Understand Risks in the Electronic Industry, *IEEE Transactions on Components, Packaging and Manufacturing Technology*, Part A, vol. *20*, pp. 542–547, Dec. 1997.

Day, J., Microcontrollers, *Electronic Buyers News*, April 10, 2000a, CMP Media, http://www.ebnews.com/story/OEG20000406S0011, accessed on January 18, 2002.

Day, J., Technology Focus: Microcontrollers, *Electronic Buyers News*, CMP Media, June 21, 2000b.

DEF-STAN-00–71, A Guide to Managing Obsolescence, DEF-STAN-00–71 Issue 1, U.K. Ministry of Defence, London, May 14, 1998.

DIN EN 62402, Deutsches Institut fuer Normung e. V. (DIN), Deutsche Norm, DIN EN 62402: 2008: Anleitung zum Obsoleszenzmanagement (IEC 62402:2007); Deutsche Fassung EN 62402:2007.

DIN ISO 9004, Deutsches Institut fuer Normung e. V. (DIN), Deutsche Norm, ISO 9004: 2000: Qualitaetsmanagementsysteme - Leitfaden zur Leistungsverbesserung. 2000.

DIN, Deutsches Institut fuer Normung e.V., About Us, http://www.din.de/cmd?level=tpl-bereich&menuid=47566&cmsareaid=47566&languageid=en, accessed on November 19, 2010a.

DIN, Deutsches Institut fuer Normung e.V., Buy Standards, http://www.din.de/cmd?workflowname=dinSearch&languageid=en, accessed on November 19, 2010b.

Directive 1907/2006/EC of the European Parliament and of the Council on the Registration, Evaluation, Authorization and Restriction of Chemicals (REACH), December 18, 2006.

Directive 2002/95/EC of the European Parliament and of the Council on the Restriction of the Use of Certain Hazardous Substances (RoHs). January 27, 2003.

Directive 2002/96/EC of the European Parliament and of the Council on Waste Electrical and Electronic Equipment (WEEE), January 27, 2003.

Defense Logistics Agency (DLA), About the Defense Logistics Agency, http://www.dla.mil/about_dla.aspx, accessed on December 5, 2010.

Easter, R., VIEWPOINT: Steer Clear of Aftermarket Parts Leprechaun, *Military & Aerospace Electronics*, May 1, 1994.

EIA-4899Standard for Preparing an Electronic Components Management Plan, EIA-4899, Electronic Industries Alliance, GEIA, Arlington, VA, n.d.

EIA-4900, Use of Semiconductor Devices Outside Manufacturers' Specified Temperature Ranges, EIA-4900, Electronic Industries Alliance, EIA, Arlington, VA, n.d.

EIA-724, Product Life-Cycle Data Model, EIA-724, Electronic Industries Alliance, Arlington, VA, September 19, 1997.

EIA, GEB1: Diminishing Manufacturing Sources and Material Shortages (DMSMS) Management Practices, EIA, http://www.geia.org/sstc/G12/geb1.html, accessed on January 14, 2002.

EIQC-0002, Component Obsolescence Standard, EIQC-0002, Electronics Industry Quality Council, November 1, 1993.

Elantec, Press Release: Elantec Semiconductor Announces Military Market Exit, Elantec Semiconductor, July 1, 1997.

EN 61025, Europaeisches Komitee fuer Normung, Europaeische Norm, EN 61025: 2007: Fehlzustandsbaumanalyse, 2007.

EN ISO 9000, Europaeisches Komitee fuer Normung, Europaeische Norm, EN ISO 9000:2000, Qualitaetsmanagementsysteme—Grundlagen und Begriffe, 2000.

FAA, CAST Position Paper P32, Federal Aviation Authority, Washington, DC, May 1, 1999.

FABTECH, Semiconductor FABTECH, 1999 Semi Copper-Critical Survey Predicts Copper-Based Semiconductors in Full Production by 2001/02, October 19, 1999, http://www.semiconductorfabtech.com/industry.news/9910/21.17.shtml, accessed on January 29, 2002.

Fairchild, Financial Reports/SEC Filings Form 10-K, Fairchild Semiconductors, November 2000, http://www.fairchildsemi.com/company/sec.pdf, accessed on April 17, 2011.

Farish, R., East Meets East, *Business Eastern Europe*, vol. *24*, pp. 17–19, October 30, 1995.

Farrell, N., Techeye.net, 64-bit Windows 7 Software on the Rise on 15.02.1010, http://www.techeye.net/hardware/64-bit-windows-7-software-on-the-rise, accessed on November 23, 2010.

Feldman, K., and Sandborn, P., Integrating Technology Obsolescence Considerations into Product Design Planning, Proceedings of the ASME 2007 International Design Engineering Conference & Computers and Information in Engineering Conference, Las Vegas, NV, September, 2007.

Feng, D., Singh, P., and Sandborn, P., *Lifetime Buy Optimization to Minimize Lifecycle Cost*, Proceedings of the 2007 Aging Aircraft Conference, 2007.

Finanzen.net, Analog Devices kann Umsatz und Gewinn deutlich steigern, 22.11.2010, http://www.finanzen.net/nachricht/aktien/Analog-Devices-kann-Umsatz-und-Gewinn-deutlich-steigern-956667, accessed on December 22, 2010.

Fitzhugh, G. L., Rapid Retargeting, A Solution to Electronic Systems Obsolescence, Proceedings of the IEEE 1998 National Aerospace and Electronics Conference (NAECON 1998), pp. 169–176, 1998.

FMEAinfocentre, Process Hazard Analysis, Failure Mode Effects Analysis (FMEA), Rev. 1, 01/15/01, http://www.fmeainfocentre.com/examples/shortintroexample.pdf, accessed on September 27, 2010.

Freescale, Life Cycle Stage Codes, Freescale Semiconductors Inc., http://www.freescale.com/webapp/sps/site/eialink.jsp, accessed on March, 20, 2010a.

Freescale, Freescale Semiconductors Inc., Support Website on Product Longevity, http://www.freescale.com/webapp/sps/site/overview.jsp?nodeId = 0570B0, accessed on October 18, 2010b.

Freescale, Freescale Semiconductors Inc., Product Website on ColdFire+ 32bit MCU, http://www.freescale.com/webapp/sps/site/homepage.jsp?code = CFPLUS, accessed on November 22, 2010c.

Freescale, Freescale Semiconductors Inc., Website on Thin Film Storage (TFS) with FelxMemory Technology, http://www.freescale.com/webapp/sps/site/overview.jsp?code = TM_RD_PROCESSTECH_90NMTFS_FLXMEM&tid = mcplusHp, accessed on November 22, 2010d.

Fyffe, S., Intel Scales Back Rambus Plans, Electronic News, August 14, 2000, Cahners Business Information, http://www.e-insite.net/electronicnews/index.asp?layout = article&articleid = CA50085&pubdate = 8/14/2000, accessed on January 15, 2002.

Garrett, B., Four Forces Shape DRAM Evolution, EE Times, February 7, 2000, CMP Media, http://www.eetimes.com/story/OEG20000204S0015, accessed on January 16, 2002.

GEIA-STD-0002–1 Aerospace Qualified Electronic Component (AQEC) Requirements, Volume 1—Integrated Circuits and Semiconductors, August 2005.

Geiger, W., Qualitaetslehre, Einfuehrung, Systematik, Terminologie. 2., voellig neubearbeitete und erweiterte Auflage, Wiesbaden, Friedr. Vieweg & Sohn Verlag, 1994.

Geiger, W., Kotte, W., Handbuch Qualitaet, Grundlagen und Elemente des Qualitaetsmanagements: Systeme Perspektiven. 5., vollstaendig ueberarbeitete und erweiterte Auflage. Wiesbaden: Friedr. Vieweg & Sohn Verlag, GWV Fachverlage GmbH. 2008.

GIDEP, Government-Industry Data Exchange Program, What Is GIDEP?, http://www.gidep.org/, accessed on April 5, 2010.

Gifford, M., High Density Designs Require a New Set of Rules for ASIC Engineers, *ASIC Technology & News*, March 1, 1990.

Giovino, William A., Microcontrollers and DSPs—Will the Two Worlds Ever Intersect?, March 2001, http://www.microcontroller.com/wp/MicrosDSPs/micros_dsps.htm, accessed on January 29, 2002.

Gowland A., Managing Software Obsolescence in the COTS Environment, Component Obsolescence Group (COG) Workshop Presentation, November 27, 2002.

Gravier, M, and Swartz S. The Dark Side of Innovation: Exploring Obsolescence and Supply Chain Evolution for Sustainment-dominated Systems, *Journal of High Technology Management Research*, vol. *20*, pp. 87–102, 2009.

Harper, C. A., *Active Electronic Component Handbook*, New York: McGraw-Hill, 1999.

Harper, C. A., Jones, H. C., John, M., and Smith, S., *Active Electronic Component Handbook*, Second Ed., New York: McGraw Hill, 1996.

Heise Online, AMD kann Mobilprozessor-Marktanteil leicht steigern, 20.08.2010, www.heise.de/newsticker/meldung/AMD-kann_Mobilprozessor-Marktanteil-leicht-steigern-1062547.html, accessed on September 21, 2010a.

Heise Online, National Semiconductors steigert Umsatz und Gewinn, 10. September 2010, http://www.heise.de/newsticker/meldung/National-Semiconductor-steigert-Umsatz-und-Gewinn-1076911.html, accessed on December 22, 2010b.

Helsen, K., and Schmittlein D., Analyzing Duration Times in Marketing: Evidence for the Effectiveness of Hazard Rate Models, *Marketing Science*, vol. *11*, pp. 395–414, 1993.

Henke A., and S. Lai, Automated Parts Obsolescence Prediction, Proceedings of the DMSMS Conference, San Antonio, TX 1997.

Herald, T. E., Technology Refreshment Strategy and Plan for Application in Military Systems: A How-to Systems Development Process and Linkage with CAIV, Proc. National Aerospace and Electronics Conference (NAECON), Dayton, OH, pp. 729–736, October 2000.

Hill, D., The Hardware Design Minefield, A Guide to the Electronic Design Process and its Obsolescence Mitigation Strategies, Component Obsolescence Group, United Kingdom, 2007.

Hitachi, SRAM Components and Modules Datasheet, Hitachi Semiconductors, July 2000.

Hitt, E. F., and Schmidt, J., Technology Obsolescence (TO) Impact on Future Costs, Proceedings of the 17th Digital Avionics Systems Conference (DASC), AIAA/IEEE/SAE, vol. *1*, pp. A33–1–A33–7, 1998.

Holzner, D., *Zur Wirtschaftlichkeit von Qualitaetsmanagementsystemen, Eine empirische Untersuchung. 01. Auflage*, Muenchen, TCW Transfer-Centrum, 2006.

Horvath, P., *Controlling. 10., vollst. ueberarb. Auflage*. Muenchen: Verlag Franz Vahlen, 2006.

Howard, M. A., *Component Obsolescence. It's Not Just for Electronics Anymore*, ARINC, Annapolis, MD, 2002.

HTV, Halbleiter-Test & Vertriebs-GmbH, Testhaus Programmiercenter Langzeitkonservierung, Company Presentation, 2009.

Huang, Z., Solomon, R., Sandborn, P., and Pecht, M., Forecasting ASIC Life-Cycles, *Future Circuits International*, vol. *7*, pp. 101–107, 2001.

Huber, J. P., and Rosneck, M.W., Successful ASIC Design the First Time Through, Van Nostrand Reinhold, 1991.

Huffman, J. O., Roadblocks in SRAM Future, *Electronic News*, Cahners Business Information, June 1998.

Humphrey, D., Condra, L., Pendse, N., Das, D., Wilkinson, C., and Pecht, M., An Avionics Guide to Uprating of Electronic Parts, *IEEE Transactions on Components and Packaging Technologies*, vol. *23*, pp. 595–599, Sept. 1, 2000.

Hynix Semiconductor, Inc., Product Overview, http://www.hynix.com/gl/products/index.jsp?menuNo = 1&m = 0&s = 0, accessed on February 11, 2010.

IEC-62239, Electronic Component Management Plans, PAS Pre-standard 62239, International Electrotechnical Committee, Geneva, Switzerland, April 1, 2001.

IEC-62240, Use of Semiconductor Devices Outside Manufacturer Specified Temperature Ranges, PAS Pre-standard 62240, International Electroctechnical Committee, April 1, 2001.

IEC-62402, IEC 62402 Ed. 1.0: Guide to Managing Obsolescence, Committee Draft (CD), International Electrotechnical Commission, IEC, 2004.

IEC-62435, International Electrotechnical Commission, IEC 62435: Electronic Components—Long-duration Storage of Electronic Components, n.d.

IEC, IEC, About the IEC, http://www.iec.ch/helpline/sitetree/about/, accessed on November 19, 2010.

Intel, 1998 Annual Report, Intel Corporation, 1998.

Intel, 1999 Annual Report, Intel Corporation, 1999.

Intel, Product Marketing Change of Embedded Pentium II Processor Low Power Module at 333 MHz and 256K L2 Cache on Die, PCN #1068–00, Intel Corporation, December 12, 2000a.

Intel, Intel Flash Memory Package Topside Lasermark Change for All EasyBGA Packages, PCN #932, Intel Corporation, February 18, 2000b.

Product Change Notification #109885–00, Selected Embeded Processor, Product Discontinuance, Intel Xeon Processor L3014, Intel Corporation, March 18, 2010a.

Intel Corporation, Moore's Law 40th Anniversary, http://www.intel.com/pressroom/kits/events/moores_law_40th/, 2010b, accessed on April, 18, 2011.

Intel Corporation, Product Website on Intel Atom Z500, http://ark.intel.com/Product.aspx?id = 35472, accessed on November 23, 2010c.

Intel Corporation, Product Website on Architecture and Silicon Technology, http://www.intel.com/technology/architecture-silicon/index.htm, accessed on November 23, 2010d.

Intel Corp., Press Kit Moore's Law 40th Anniversary, http://www.intel.com/pressroom/kits/events/moores_law_40th/index.htm, accessed on November 30, 2010e.

Intel Corp., Information on Technology Website, http://www.intel.com/technology/manufacturing/index.htm?iid = subhdr-DE + tech_mfg, accessed on December 26, 2010f.

IT-Times, Texas Instruments blickt zuversichtlich ins neue Jahr. hohe Nachfrage nach Smartphone-Chips sorgt fuer gute Geschaefte, http://www.it-times.de/news/nachricht/datum/2010/12/08/texas-instruments-blickt-zuversichtlich-ins-neue-jahr-hohe-nachfrage-nach-smartphone-chips-sorgt-f/?cHash = 6959feed61&type = 98, accessed on December 21, 2010.

Jackson, M., Sandborn, P., Pecht, M., Hemens-Davis, C., and Audette, P., A Risk-Informed Methodology for Parts Selection and Management, *Quality and Reliability Engineering International*, vol. *15*, pp. 261–271, 1999.

JESD46A, *Guidelines for User Notification of Product/Process Changes by Semiconductor Suppliers*, JESD46A, Electronic Industries Alliance, Arlington, VA, April 1, 1994.

JESD48, *Product Discontinuance*, JESD48, Electronics Industry Association, Arlington, VA, September 1, 1997.

JESD53, *Management of Component Obsolescence by Government Contractors*, JESD53, Electronics Industry Association, Arlington, VA, January 1, 1996.

John, M., and Smith, S., *Application-Specific Integrated Circuits*, Addison-Wesley, 1997.

Johnson, G., Flash Memory Sales Soar, *Electronic News*, June 19, 2000, Cahners Business Information, http://www.e-insite.net/electronicnews/index.asp?layout=article&articleid=CA49672&pubdate=6/19/00, accessed on January 14, 2002.

Jones, A., The Emulation and Substitution Minefield, Guide to Replacing Obsolete Semiconductor Components, Component Obsolescence Group, United Kingdom, 2008.

Josias, C., and Terpenny, J., Component Obsolescence Risk Assessment, Proceedings of the Industrial Engineering Research Conference (IERC), Houston, TX 2004.

Kallender, P., NEC Talking Global, *Electronic News*, July 14, 2000, Cahners Business Information.

Karls, J., Dickens, H., and Shon-Roy, L., *Status 1998: A Report on the Integrated Circuit Industry*, ICE, Scottsdale, AZ, 1998.

Kemet Electronic Cooperation, Ceramic High Reliability General Information GR900 Series, http://www.kemet.com/kemet/web/homepage/kechome.nsf/weben/B1437EF3A2EC44B2852571A2005A3325/$file/GR900CE.pdf, accessed on August 27, 2010.

Kersten, G., Handbuch Qualitaetsmanagement, Fehlermoeglichkeits-und-einflussanaylse (FMEA), 3., gruendliche ueberarbeitete und erweiterte Auflage, Muenchen, Carl Hanser-Verlag (published by Masing, W.), 1994.

Kline, J., Outsourcing in the Electronics Industry, *HDI Magazine*, vol. *4*, no. 2, pp. 40–42, February 1, 2001.

KPMG International, Electronics, Software & Services, Managing the Risks of Counterfeiting in the Information Technology Industry, 2005.

Kroeger, R., Manufacturer Part Assessment: A Pragmatic Approach, First Workshop on Use of Components Outside Specified Temperature Limits, CALCE EPSC University of Maryland, Jan. 1997.

Kuehn, L. S., P-3 Program Office Component Obsolescence Planning Guide, Prepared for Naval Air Systems Command by ARINC Inc., November 1984.

LePedus, EE Times, News & Analysis Website, Intel Gains MPU Share in Slow Q1, Le Pedus, M., April 29, 2010, http://www.eetimes.com/electronics-news/4088739/Intel-gains-MPU-share-in-slow-Q1, accessed on November 23, 2010.

Levitt, T., Exploit the Product Life Cycle, *The Harvard Business Review*, vol. *43*, pp. 81, 14p, Nov. 1965.

Lineback, J. R., Two Good Years May Move up DRAM Sales to Peak '95 Level, Semiconductor Business News, January 3, 2000, CMP Media, http://www.siliconstrategies.com/story/OEG20000103S0021, accessed on January 18, 2002.

Livingston, H., SSB-1: Guidelines for Using Plastic Encapsulated Microcircuits and Semiconductors in Military, Aerospace and Other Rugged Applications, Defense Microelectronics Activity, December 14, 2000, http://www.dmea.osd.mil/ssb1_paper.pdf, accessed on January 15, 2002.

Lukasiewicz, J., *The Railway Game*, Ontario, Canada: McClelland and Stewart Limited, 1976.

macnews.de, Mac-Verkaeufe: Starkes Wachstum in Behoerden und Unternehmen, Florian Matthey, http://www.macnews.de/news/51874/mac-verkaufe-starkes-wachstum-in-behorden-und-unternehmen/, accessed on October 15, 2010.

Makowski, D., Winhistory Website by Dipl. Inf. (FH) http://www.winhistory.de/more/timeline.htm, accessed on October 4, 2010.

Makowski, D., Winhistory Website by Dipl. Inf. (FH) http://www.winhistory.de/more/386/1400vers.htm, accessed on October 12, 2010.

Mann, C. C., The End of Moore's Law?, *Technology Review*, May 1, 2000, http://www.techreview.com/articles/mann0500.asp, accessed on January 15, 2002.

Markt & Technik, Weniger Geld mehr Elektronik, Markt & Technik, Edition 14/2010a.

Markt & Technik, Rekordgewinne fuer Halbleiterhersteller, Markt & Technik, Edition 13/2010b.

Martino, J. P., *Technological Forecasting for Decision Making*, Second Edition, North-Holland, 1983.

Mayer, J. H., Buying Strategies: Embedded Processors, *Electronic Buyers News*, March 6, 2000, CMP Media, http://www.ebnews.com/story/OEG20000302S0014, accessed on January 15, 2002.

McAdam, R., The Future of Digital Is Analog, *Electronic News*, August 16, 1999, Cahners Business Information, http://www.e-insite.net/electronicnews/index.asp?layout=article&articleid=CA66896&pubdate=8/16/1999, accessed on January 15, 2002.

The McClean Report, McClean, B., Matas, B., Yancey, N., and Yancey, T., IC Insights, Scottsdale, AZ, 2007.

McCray, K., Are Individual Domestic Test and Assembly Houses in for a Change?, *HDI Magazine*, vol. *4*, no. 2, pp. 20–21, February 1, 2001, http://www.hdi-online.com/mag/2001/02/0102dip.html, accessed on January 29, 2002.

McDaniel, G., *IBM Dictionary of Computing*, New York: McGraw-Hill, 1993.

McDermott, J., Shearer, J., Tomczykowski, W., Resolution Cost Factors for Diminishing Manufacturing Sources And Material Shortages, *Defense Microelectronics Activity (DMEA)*, 1999.

McIlvaine, B., Microprocessors: 2000 Will Be More of Same, No Big Surprises, *Semiconductor Business News*, January 1, 2000, CMP Media, http://www.siliconstrategies.com/story/OEG20000103S0022, accessed on January 14, 2002.

Memphis Memory Distribution AG, Product Website on DRAM, http://www.memphis.ag/german/edo_fpm_dram.html, accessed on October 23, 2010a.

Memphis Memory Distribution AG, Product Website on SRAM, http://www.memphis.ag/german/sram.html, accessed on November 1, 2010b.

Merola, L., The COTS Software Obsolescence Threat, Proceedings of the International Conference on Commercial-off-The-Shelf (COTS) Based Software Systems, 2006.

Micron, Technical Note TN-00-18 Rev. E: Temperature Uprating on Semiconductors, Micron Semiconductors, Inc., May 2008, http://download.micron.com/pdf/technotes/TN0018.pdf., accessed on March 20, 2010.

Micron, Product Page, http://www.micron.com/products/partdetail?part=MT46H128M16LFCK-5%20IT, accessed on April 19, 2010a.

Micron, Product Page, http://www.micron.com/products/partdetail?part=MT42L64M64D4KJ-3%20IT, accessed on April 19, 2010b.

Micron, Technology, http://numonyx.com/en-US/MemoryProducts/NOR/Pages/P30P33.aspx, accessed on September 7, 2010c.

Microsoft, Microsoft Press Service, Microsoft Inc, http://www.microsoft.com/germany/presseservice/news/pressemappe.mspx?id = 532216, accessed on October 2, 2010.

Microsoft, Microsoft Support Page, Microsoft Inc., http://support.microsoft.com/gp/lifepolicy, accessed on October 2, 2010b.

Microsoft, Microsoft Life-cycle Support Page, Microsoft Inc., http://support.microsoft.com/lifecycle/?LN = en-gb&C2 = 1173, accessed on October 4, 2010c.

Microsoft, Microsoft Windows 7 Product Page, Microsoft Inc., System Requirements, http://www.microsoft.com/windows/windows-7/get/system-requirements.aspx, accessed on October 11, 2010d.

Microsoft, Microsoft Help and Support Page, Microsoft Inc., System Requirements Windows XP Professional, http://support.microsoft.com/kb/314865, accessed on October 11, 2010.

Microsoft, Microsoft Inc., Support Website, Life-cycle Policy, http://support.microsoft.com/gp/lifepolicy, accessed on October 15, 2010f.

MIL-PRF-19500M, Semiconductor Devices: General Specification for MIL-PRF-19500M, U.S. Department of Defense, October 1, 1999.

MIL-PRF-38534D, Hybrid Microcircuits: General Specification for, MIL-PRF-38534D, U.S. Department of Defense, January 1, 1999.

MIL-PRF-38535, Test Methods and Procedures for Microelectronics, MIL-PRF-38535, U.S. Department of Defense, November 1, 1989.

Miller F. P., Vandome, A. F., McBrewster, J. (Ed.), *History of Microsoft Windows*, Alphascript Publishing, 2009a.

Miller, F. P., Vandome, A. F., McBrester, J. (Ed.), *History of Apple*, Alphascript Publishing, p. 52, 2009.

Milwaukee Art Museum, Industrial Strength Design, How Brooks Stevens Shaped Your World, Milkwaukee Art Museum Biography, http://www.mam.org/collection/archives/brooks/index.asp, accessed on April 5, 2010.

Mittelstaedt, T., Network Community, Managing Software (and System) Obsolescence, *Computer Bits*, vol. 9, November 1999.

Motorola, Motorola Continues as Industry Leader of Worldwide Microcontroller Market, Press Release, November 30, 1999, http://www.apspg.com/press/press/113099/microcontroller.html, accessed on January 18, 2002.

MSC Vertriebs GmbH, SRAM Speicher Website, http://www.msc-ge.com/de/produkte/elekom/g1/sram/2346-www.html?pos = 0, accessed on November 1, 2010.

Muessigmann, N., Strategische Liefernetze, 1. Auflage. Wiesbaden, Deutscher Universitaets-Verlag, GWV Fachverlage GmbH, 2007.

Murphy, T., DRAM Advances Chip Growth, Electronic News, January 10, 2000a, Cahners Business Information, http://www.e-insite.net/electronicnews/index.asp?layout = article&articleid = CA48639&pubdate = 1/10/2000, accessed on January 15, 2002.

Murphy, T., Toshiba, Hitachi fall: Will others follow?, *Electronic News*, June 26, 2000b, Cahners Business Information, http://www.e-insite.net/electronicnews/index.asp?layout = article&articleid = CA49718&pubdate = 6/26/2000, accessed on January 15, 2002.

Murphy, T., Cornering the Flash Market, Electronic News, March 6, 2000, Cahners Business Information, http://www.e-insite.net/electronicnews/index.asp?layout=article&articleid=CA48977&pubdate=3/6/00, accessed on January 14, 2002.

National Semiconductor, Product Folder, LF11331 - 4 Normally Open Switches with Disable [Discontinued], January 14, 2002a, http://www.national.com/pf/LF/LF11331.html, accessed on January 14, 2002.

National Semiconductor, Change Management System, http://www.national.com/quality/0,1790,90,00.html, accessed on January 15, 2002b.

NEDA, We're Established, National Electronic Distributors Association, 1999, http://nedafoundation.org/industry/index.html, accessed on January 15, 2002.

Nielsen, A. P., Outsourcing and the Development of Competencies, PICMET'99, Portland, OR, pp. 72–77, July 1999.

Nelson III, R. and Sandborn, P., Strategic Management of Component Obsolescence Using Constraint-Driven Design Refresh Planning, to be published *International Journal of Product Life Cycle Management*, 2012.

Okey, D. W., Products Liability and Uprating of Electronic Components, *Journal of Air Law and Commerce*, pp. 1201–1255, Fall 1999.

Olfert, K., Projektmanagement, Kompakt-Training. 6. Auflage. Ludwigshafen (Rhein), Friedrich Kiehl Verlag, 2008.

ON Semiconductor, New Mask Set for LM317T/D2T from Tesla, Change Notification #10001, ON Semiconductor, August 28, 1999.

ON Semiconductor, Assembly/Test Site Change from Motorola KLM t Seznam Czech Republic, Change Notification #10135, ON Semiconductor, February 10, 2000a.

ON Semiconductor, Seefull China Fab/Assy Site Qual for General Purpose Rectifiers, Change Notification #10340, ON Semiconductor, October 24, 2000b.

ON Semiconductor, ON Semiconductors, Datasheet on NCV8570, http://www.onsemi.com/pub_link/Collateral/NCV8570-D.PDF, accessed on December 21, 2010.

Orbach, Barak Y., The Durapolist Puzzle: Monopoly Power in Durable-Goods Markets, *Yale Journal on Regulation* vol. *21*, pp. 67–118, 2004.

Oswalt, D., and Andrey, D., Global Semiconductor Market to Grow 31% in 2000 Says SIA, Semiconductor FABTECH.com, June 8, 2000, http://www.fabtech.org/industry.news/0006/08.04.shtml, accessed on January 18, 2002.

Pabst, T., Intel Roadmap Step by Step, Tom's Hardware Guide, February 8, 2001, http://www6.tomshardware.com/cpu/01q1/010208/roadmap-06.html, accessed on January 14, 2002.

Palmer, M. A., Alternative Processing for Forming Solder Joints with Lead-Free Solder Alloys, Surface Mount Technology, August 2000, PennWell Corporation, http://smt.pennnet.com/Articles/Article_Display.cfm?Section=Archives&Subsection=Display&ARTICLE_ID=80845, accessed on January 15, 2002.

Pecht, M., and Das, D., The Electronic Part Life-cycle, *IEEE Transactions on Components and Packaging Technologies*, vol. *23*, pp. 190–193, March 2000.

Pecht, M., *Plastic Encapsulated Microelectronics*, New York: John Wiley & Sons, 1995.

Pecht, M., Fink, J., Hakim, E., and Wyler, J., An Assessment of the Qualified Manufacturer List (QML), *IEEE Aerospace and Electronic Systems Magazine*, vol. *12*, pp. 39–43, July 1, 1997.

Pecht, M., Solomon, R., Sandborn, P., and Wilkinson, C., *Life Cycle Forecasting, Mitigation Assessment and Obsolescence Strategies*, CALCE EPSC Press, 2002.

Pennington, S., Before We Improve It... We've Got to Prove It: The Role of the Technical Review Board, *IBM Microelectronics*, First Quarter, pp. 32–34, 1999.

Perry, W. J., Secretary of Defense, press conference, June 29, 1994.

Philips Semiconductors, Philips Semiconductors Is the First to Create Cost Effective Embedded Flash Memory Process for 0.18 Micron CMOS, Phillips press release, January 24, 2000, http://www.semiconductors.philips.com/news/content/file_511.html, accessed on January 18, 2002.

Phillips, C., Controlling and Managing Engineering Changes, *Manufacturing Systems*, *5*, pp. 40–43, December 1, 1987.

Pleyma, Material-Rundbrief, Mai 2010, http://www.pleyma.de/pleyma/fileadmin/Downloads/Rundbrief/2010_05_04_Rundbrief_Mai_PLEYMA.pdf, accessed on May 16, 2010.

Porter, Z. An Economic Method for Evaluating Electronic Component Obsolescence Solutions, Seattle, WA, May 1, 1998, http://www.geia.org/sstc/G12/porter_paper.pdf, accessed on January 18, 2002.

Prabhakar, V., and Sandborn, P., A Part Total Ownership Cost Model for Long-Life Cycle Electronic Systems, *International J. of Computer Integrated Manufacturing*, vol. *24*, 2011.

Predko, M., *Handbook of Microcontrollers*, New York: McGraw-Hill Inc., 1999.

Prince, B., *High Performance Memories*, West Sussex, England: John Wiley and Sons, 1999.

Qin, University of Berkeley, Electrical Engineering and Computer Sciences Department, Research work on SRAM by Qin, H., https://buffy.eecs.berkeley.edu/PHP/resabs/resabs.php?f_year=2006&f_submit=one&f_absid=101068, accessed on November 1, 2010.

QinetiQ and ARINC, QinetiQ and ARINC, Ministry of Defence Component Obsolescence Resolution Cost Metrics Study, United Kingdom Ministry of Defence (UK MoD), 2004.

QTEC, 2006QTEC/QinetiQ, http://www.qtec.us/Products/QStar_Introduction.htm, accessed on January 15, 2002.

Rai, R., and Terpenny, J., Principles for Managing Technological Product Obsolescence, *IEEE Transactions on Components and Packaging*, vol. *31*, pp. 880–889, 2008.

ReliaSoft's Xfmea Software, Xfmea Report Sample Design FMEA, Created 3/26/2003, http://www.fmeainfocentre.com/examples/xfmea_dfmea.pdf, accessed on September 27, 2010.

Rentrop, Netzwelt.de, Magazin fuer IT& Consumer Electronics, Apple + Intel: Sie habens getan by Rentrop, C., http://www.netzwelt.de/news/71465_2-apple-intel-habens-getan.html, accessed on November 23, 2010.

Rickman T., and Singh, G., Strategies for Handling Obsolescence, End-of-Life and Long-Term Support of COTS Software, *COTS Journal*, pp. 17–21, January 2002.

Riemenschneider, Elektroniknet.de, Renesas dominiert Mikrocontroller, Infineon neue Nr. 1 im Automotive-Geschaeft, by Riemenschneider, F., http://www.elektroniknet.de/automotive/technik-know-how/bauelemente/article/26993/0/Renesas_dominiert_Mikrocontroller_Infineon_neue_Nr1_im_Automotive-Geschaeft/, accessed on December 21, 2010a.

Riemenschneider, Elektroniknet.de, Die groesssten Halbleiterhersteller 2010, by Riemenschneider F., http://www.elektroniknet.de/bauelemente/news/article/31350/0/Die_groessten_Halbleiterhersteller_2010/, accessed on December 22, 2010b.

Ristelhueber, Ristelhueber, R., SIA Sees Stronger Growth for Chip Market in 2000, EETimes, October 1, 1999, CMP Media, http://www.csdmag.com/story/OEG19991027S0048, accessed on January 18, 2002.

Roberts, B., No Strategy Worth a DRAM, Electronic Business, February 1, 1999, Cahners Business Information, http://www.e-insite.net/eb-mag/index.asp?layout=article&stt=001&articleid=CA66370&pubdate=2/1/1999, accessed on January 15, 2002.

Robins, M., The History of Pb-Free, Electronic Packaging and Production, June 1, 2000, Cahners Business Information, http://www.e-insite.net/epp/index.asp?layout=article&stt=001&articleid=CA73740&pubdate=6/1/00, accessed on January 15, 2002.

Rochester Electronics, Defending Against Counterfeit Electronic Parts and Gray Market Practices, Rochester Electronics, 2007.

Rochester Electronics, about Rochester Electronics, http://www.rocelec.com/about/about_rochester_electronics/, accessed on April 19, 2010.

Rogokowski, R., The Supply Chain Minefield, The Role of the Distributor in Managing Obsolescence Problems, Component Obsolescence Group, United Kingdom, 2003.

Rogokowski, R., The Obsolescence Minefield, A Guide for Senior Executives, Issue 1, Component Obsolescence Group, United Kingdom, 2007.

Rogowski, R., UK Electronics Alliance, UKEA Position on Counterfeit Electronic Components, RR/V2/03.03.2008.

Ross, G., Electronics Advocate, 08.04.2010, iPad BOM is $259.60, says Issuply, http://www.electronicsadvocate.com/2010/04/08/ipad-bom-is-25960-says-isuppli/, accessed on December 21, 2010.

Samsung, SRAM products data sheets, Samsung Semiconductors, July 2000.

Samsung, Samsung Electronics Customer Support, http://www.samsungelectronics.com/semiconductors/qa/customer_support/customer_support_p1.html, accessed on January 18, 2002.

Samsung Semiconductor, Inc, DS-09-SRAM-001, Samsung QDR II SRAM- High-Bandwidth Memory for Advanced Network Equipment, Printed 01/2009, http://www.samsung.com/global/business/semiconductor/support/brochures/downloads/memory/qdr2_datasheet_200902.pdf, accessed on November 1, 2010.

Samsung Semiconductor, DRAM Product Website, http://www.samsung.com/global/business/semiconductor/products/dram/Products_ComputingDRAM.html, accessed on September 25, 2010a.

Samsung Semiconductor, Inc., Product Website on High Speed SRAM, http://www.samsung.com/global/business/semiconductor/products/sram/Products_HighSpeedSRAM.html, accessed on November 1, 2010b.

Samsung Semiconductor, Inc., Product Website on Computing DDR3 DRAM, http://www.samsung.com/global/business/semiconductor/productInfo.do?fmly_id=691&partnum=K4B4G0846A&xFmly_id=690, accessed on November 2, 2010c.

Samsung Semiconductor Inc., Press Releases SAMSUNG Now Producing 20nm-class, 64-gigabit 3bit NAND Flash Memory, Seoul, Korea, Oct. 13, 2010, http://www

.samsung.com/global/business/semiconductor/newsView.do?news_id=1207, accessed on November 22, 2010d.

Sandborn, P., Software Obsolescence: Complicating the Part and Technology Obsolescence Management Problem, *IEEE Trans on Components and Packaging Technologies*, vol. *30*, no. 4, pp. 886–888, December 2007.

Sandborn, P., Strategic Management of DMSMS in Systems, *DSP Journal*, pp. 24–30, April/June 2008.

Sandborn, P. A., Cost Analysis of Electronic Systems, *World Scientific*, 2011.

Sandborn, P., Herald, T., Houston, J., and Singh, P. Optimum Technology Insertion into Systems Based on the Assessment of Viability. *IEEE Trans. on Comp. and Pack. Tech*, *26*, pp. 734–738, 2003.

Sandborn, P., Jung R., Wong R., and Becker J., A Taxonomy and Evaluation Criteria for DMSMS Tools, Databases and Services, Proceedings of the 2007 Aging Aircraft Conference, Palm Springs, CA, April 2007.

Sandborn, P., Mauro, F., and Knox, R., *A Data Mining Based Approach to Electronic Part Obsolescence Forecasting*, IEEE Trans. on Components and Packaging Technology, vol. *30*, pp. 397–401, September 2007.

Sandborn, P., Prabhakar, V., and Feng, D., DMSMS Lifetime Buy Characterization via Data Mining of Historical Buys, Proceedings DMSMS Conference, Orlando, FL, November 2007.

Sandborn, P., Prabhakar, V., and Ahmad, O., Forecasting Electronic Part Procurement Lifetimes to Enable the Management of DMSMS Obsolescence, *Microelectronics Reliability*, vol. *51*, pp. 392–399, 2011.

SAP Online Help, SAP Library, SAP Business One, http://help.sap.com/saphelp_sbo2005asp1/helpdata/de/2f/fa1e5515517848baf7c742f8e56e18/content.htm, accessed on September 28, 2010.

Schroeter, J., *Surviving the ASIC Experience*, New York: Prentice Hall, 1992.

SD-22, Defense Standardization Program Office, SD-22 Diminishing Manufacturing Sources and Material Shortages, A Guidebook of Best Practices and Tools for Implementing a Proactive DMSMS Management Program, September 2010, http://dksp.dcnicn.com/files/SD-22_DMSMS_Guidebook_Update_09-2010.pdf, accessed on December 5, 2010.

Seghezzi, H. D., Integriertes Qualitaetsmanagement. Das St. Galler Konzept, Muenchen, Carl Hanser-Verlag, 1996.

Sematech, International Sematech, International Sematech Official Dictionary, Rev 5.0, http://www.sematech.org/public/publications/dict/index.htm, accessed on January 14, 2002.

Shaffer, G., and McPherson, G., FAA COTS Risk Mitigation Guide: Practical Methods for Effective COTS Acquisition and Lifecycle Support, Revision 3.2, January 2010.

Sharma, K. A., *Semiconductor Memories*, Piscataway, NJ: IEEE Press, 1997.

Shaw, W., Speyerer, F., Sandborn, P., Diminishing Manufacturing Sources and Material Shortages (DMSMS), Non-Recurring Engineering (NRE) Cost Metric Update, Prepared for Defense Microelectronics Activity, McClellan, September 2010, http://www.bis.doc.gov/defenseindustrialbaseprograms/osies/defmarketresearchrpts/cost_metric_final.pdf, accessed on December 6, 2010.

Shearer, J., and Tomczykowski, W., Resolution Cost Metrics for Diminishing Manufacturing Sources and Material Shortages, Supplemental Report, Defense Microelectronics Activity (DMEA), 2001.

Sheppard R., *Software Obsolescence*, COG Workshop Presentation, May 19, 2005.

SIA, White Section: SIA/GPC Perspective and Recommendations on DMS, Semiconductor Industries Association, 1996, http://www.semichips.org/, accessed on January 15, 2002.

SIA, International Technology Roadmap for Semiconductors, Semiconductor Industries Association, 1999.

Singh, P., and Sandborn, P., Obsolescence Driven Design Refresh Planning for Sustainment-dominated Systems, *The Engineering Economist*, vol. *51*, pp. 115–139, April-June 2006.

Singh, P., Sandborn, P., Lorenson, D., and Geiser, T., Determining Optimum Redesign Plans for Avionics Based on Electronic Part Obsolescence Forecasts, Proceedings of the International Conference on Concurrent Engineering, pp. 961–970, July 2002.

Slade, G., *Made to Break: Technology and Obsolescence in America*, Harvard University Press, 2007.

Smith, W. J., *DESC DMSMS Workshop Minutes, Prepared for DESC*, September 20–22, 1983. Defense Supply Center, Columbus, OH.

Smith, T., Defence Evaluation and Research Agency, Whole Life Support Department, Future Initiatives for Obsolescence Mitigation Strategies, n.d.

Solomon, R., Sandborn, P., and Pecht, M., Electronic Part Life-cycle Concepts and Obsolescence Forecasting, *IEEE Transactions on Components and Packaging Technologies*, vol. *23*, pp. 707–717, Dec. 1, 2000.

Sondermann, J. P., Handbuch Qualitaetsmanagement, Interne Anforderungen und Anforderungsbewertung. 3., gruendliche ueberarbeitete und erweiterte Auflage. Muenchen: Carl Hanser-Verlag (hrsg. von Masing, Walter). 1994.

STACK, General Requirements for Integrated Circuits, S0001 Issue 12.2, Notice 2, STACK International, St. Albans, UK, November 1, 1999.

STD-0002-1, ANSI/GEIA STD-0002-1, Aerospace Qualified Electronic Component (AQEC) Requirements, Volume 1—Integrated Circuits and Semiconductors.

Stogdill, R. C., Dealing with Obsolete Parts, *IEEE Design & Test of Computers*, vol. *16*, pp. 17–25, April 1999.

Strauss, W., DSP Market Update, Zilog, May 1, 2000, http://www.zilog.com/about/relatednews/030600.html, accessed on January 15, 2002.

Sullivan, L., Going Global: Around the World Everywhere the Same, Electronic Buyers News, May 17, 2001, CMP Media, http://www.ebnews.com/story/OEG2001 0511S0027, Accessed on January 21, 2001.

Syrus, T., Ramgopal, U., and Pecht, M., Manufacturer Assessment Procedure and Criteria for Parts Selection and Management, *IEEE Transactions on Electronics Packaging and Manufacturing*, vol. *24*, pp. 351–358, Oct. 2001a.

Syrus, T., Pecht, M., and Humphrey, D., Part Assessment Guidelines and Criteria for Parts Selection and Management, *IEEE Transactions on Electronics Packaging Manufacturing*, vol. *24*, pp. 339–350, Oct. 2001b.

Telleen-Lawton, D., Private Communication, Cogent Software Inc., 1999 (UnPub).

Texas Instruments, 1999 Standard Linear and Logic Product Withdrawal Device List, Texas Instruments, September 2, 1999a.

Texas Instruments, Qualification Notification for Select SN74CBT Technology Devices, PCN: 199990624001, Texas Instruments, August 30, 1999b.

Texas Instruments, Notification of Assembly Material Changes in the TO-220 Package Built at PSI Technologies Inc. in Manila, Philippines, PCN: 19990623003, Texas Instruments, November 9, 1999c.

Texas Instruments, Informational Notification for the Removal of Texas Instruments Logo and Texas Instruments Name on Shipping Materials, PCN: 19990120001, Texas Instruments, February 23, 1999d.

Texas Instruments, All Sites/All SBE's: Universal Change to the Box/Bag/Reel Label, PCN: 19990401001, Texas Instruments, May 19, 1999e.

Texas Instruments, Military products: TI Military Lifetime Buy Notices, Texas Instruments, December 31, 1999f.

Texas Instruments, Qualification Notification for the SN65ALS180 and SN75ALS180 Die Revision K, PCN: 19991105003, Texas Instruments, November 17, 2000a.

Texas Instruments, Texas Instruments HATO Wafer Fab. Closing: Lifetime Buy Notice (ASIC), PCN: 19990908001, Texas Instruments, June 16, 2000b.

Texas Instruments, Qualification Notification for SN74LVC161284 Device Redesign, PCN: 20000414001, Texas Instruments, June 7, 2000c.

Texas Instruments, Qualification Notification for the SN65ALS176, SN75ALS176, SN75ALS176A, SN75ALS176B, SN65ALS1176, SN75ALS176, SN75ALS176A, SN75ALS176B, SN65ALS1176, PCN: 19991216004, Texas Instruments, January 7, 2000d.

Texas Instruments, Qualification of the Sumitomo EME5050 Mold Compound, PCN: 20000418001, Texas Instruments, December 20, 2000e.

Texas Instruments, Informational Notification of Quantity Change for Select Devices Previously Offered by Harris, PCN: 20000202006, Texas Instruments, February 17, 2000f.

Texas Instruments, Texas Instruments, First Half 2001 Logic Selection Guide—10th Anniversary Edition, November 2000, http://www.ti.com/sc/docs/general/logic/lit/intro.htm, accessed on January 15, 2002g.

Texas Instruments, Texas Instruments, Product/Process Change Notifications (PCN), https://mist.ext.ti.com/pcn, accessed on January 14, 2001a.

Texas Instruments, Qualification of 5DBV Package at Hana, PCN: 20000606002, Texas Instruments, February 13, 2001b.

Texas Instruments, HIJI/TITL MicroStar Junior BGA Solder Ball Size Change, PCN: 20010817000, Texas Instruments, November 30, 2001c.

Texas Instruments, Notification of Additional Lead Finish, PCN: 20000718001, Texas Instruments, March 29, 2001d.

Texas Instruments Inc., Product Website on SN74BCT374DBR, Datasheet Revised March 2003, http://focus.ti.com/lit/ds/symlink/sn54bct374.pdf, accessed on December 22, 2010.

Texas Instruments Incorperated, Interactive Timeline of Innovation, http://www.ti.com/corp/docs/company/history/interactivetimeline.shtml, accessed on November 23, 2010a.

Texas Instruments Inc., Product Website on LM3S9B96, Datasheet Revised December 2010, http://focus.ti.com/lit/ds/symlink/lm3s9b96.pdf, accessed on December 22, 2010b.

Thommen, J.-P., and Achleitner, A.-K., Allgemeine Betriebswirtschaftslehre, Umfassende Einfuehrung aus managementorientierter Sicht, 4., ueberarb. und erw. Auflage, Wiesbaden, Gabler Verlag, 2005.

Tomczykowski, W., DMSMS Acquisition Guidelines. Implementing Parts Obsolescence Management Contractual Requirements Rev. 3.0, Defense Microelectronics Activity (DMEA), 2001.

Tomczykowski, W., Fritz, A., and Scalia, R., Program Managers Handbook. Common Practices to Mitigate the Risk of Obsolescence. Revision D, Prepared for Defense Microelectronics Activity (DMEA), 2000.

Top Statistics Index, Semiconductor Electronics Resource Center, http://www.militaryelectronics.com/topstatistics.htm, accessed on January 15, 2002.

Toronto Star, 25th Anniversary for Microprocessor, Toronto Star and Fast Forward, November 17, 1996, http://www.computerwriter.com/archives/1996/sf171196.htm, accessed on January 14, 2002.

Toshiba, Memory Products SRAM Data Sheets, Toshiba America Electronic Components Inc., July 2000.

Toshiba, Toshiba Semiconductor, Inc., Introduction to Company Profile Website, http://www.semicon.toshiba.co.jp/eng/profile/division/index.html, accessed on November 15, 2010a.

Toshiba, Toshiba Corporation, News Release/Topics, Toshiba Launches 24 nm Process NAND Flash Memory, released on Aug 31, 2010, http://www.semicon.toshiba.co.jp/eng/event/news/index.html, accessed on November 23, 2010.

Tyco Electronics, Product Details of CPF0603F100RC1, http://www.tycoelectronics.com/catalog/pn/en/6-1879335-5?RQPN=CPF0603F100RC1, accessed on September 28, 2010.

U.S. DoD, Mandatory Procedures for MDAPs and MAIS Acquisition Programs, Draft DoD 5000.2-R, U.S. Department of Defense, April 1, 2000.

U.S. DoD, Air Force Manufacturing Technology 2001 Project Book, U.S. Department of Defense, Air Force Research Laboratories Wright Patterson Air Force Base, Dayton, OH, 2001a.

U.S. DoD, Case Resolution Guide, U.S. Department of Defense Air Force Materiel Command, Universal Technology Corporation (UTC), Dayton, OH, March 31, 2001b, http://web1.deskbook.osd.mil/, accessed on January 18, 2002.

U.S. DoD, Joint Technical Architecture V3.0, U.S. Department of Defense, http://www-jta.itsi.disa.mil/jta/jtav3-final-19991115/finalv3.html, accessed on January 15, 2002.

U.S. DoD, U.S. Department of Defense, Defense Acquisition Deskbook. http://web2.deskbook.osd.mil/, accessed on January 15, 2002.

U.S. DoD, U.S. Department of Defense (DoD), Diminishing Manufacturing Sources and Material Shortages (DMSMS) Guidebook, March 20, 2005.

U.S. DoD, U.S. Code: Title 10, Chapter 131, Section 2213, Limitation on Acquisition of Excess Supplies, U.S. Department of Defense, http://www4.law.cornell.edu/uscode/10/2213.html, accessed on April 17, 2011.

USADATA, Worldwide Microcontroller Market, August 1, 1999, http://www.usadata.com/, accessed on January 18, 2002.

Usenix, the Advances Computing Systems Association, Peter Gutmann, IBM T.J: Watson Research Center, Data Remanence in Semiconductor Devices, http://www.usenix.org/events/sec01/full_papers/gutmann/gutmann_html/, accessed on November 1, 2010.

VDE, Verband der deutschen Elektrotechnik Elektronik und Informationstechnik e.V. (VDE), Pressemitteilung VDE-Studie sieht fuer Elektroautos noch Huerden bis zum Massenmarkt, http://www.vde.de/de/Verband/Pressecenter/Pressemeldungen/Fach-und-Wirtschaftspresse/Seiten/2010–35.aspx, accessed on April 28, 2010.

Viking Components Inc., Viking Components Expands SRAM Offerings to 16 MB, Viking Press Release, July 12, 2000, http://www.vikingcomponents.com/news/releases/sram.cfm, accessed on January 16, 2002.

Vishay Vitramon, Datasheet, High Rel C0G (NPO), Document number: 45045, Revision 26.02.2010.

VITA, VMEbus International Trade Association, VITA 53.0 Standard for Commercial Technology Market Surveillance, VITA Standards Organisation Approved on November 17, 2009.

Walden, D., Blanshan, K., and Kranz, G., Technology Roll Lessons Learned on Embedded Avionics Platforms, *IEEE Aerospace and Electronics Systems Magazine*, vol. *15*, pp. 19–23, May 1, 2000.

Walsh, P., Technology Focus: Torpid SRAM Market Gets a Needed Jolt, Electronic Buyers News, July 1, 1999, CMP Media, http://www.ebnews.com/story/OEG19990701S0011, accessed on January 15, 2002.

Wasson, J., and Wilkinson, C., Component Obsolescence Management for Military and Space Electronics, 5th Annual International Conference, Conference & Exhibition, Commercialization of Military and Space Electronics, Los Angeles, Feb. 2001.

Weber, P., Making Waves in the DSP Market, Soloman Brothers Conference, New York, September 10, 1997, http://www.ti.com/corp/docs/investor/speeches/salomon.shtml, accessed on January 18, 2002.

Wiersch, B., Die Kaefer-Chronik, Die Geschichte einer Autolegende, DeliusKlasing, Bielefeld 2005.

Wilkinson, C., Wasson, J., Condra, L., Fink, J., Scheidt, C., Brennom, T., Porter, Z., Chopin, J-M., Hoad, R., Matthews, J., Chapman, J., Humphrey, D., Feussinger, P., and Sagot, Y., Component Obsolescence Management for Aerospace Electronic Equipment, Proceedings of 1999 Avionics Conference, Civil and Military Convergence, London (Heathrow), UK, Nov. 1999.

Wright, M., Humphrey, D., and McCluskey, P., Uprating Electronic Components for Use Outside Their Temperature Specification Limits, *IEEE Trans. on Components, Packaging and Manufacturing Technology*, Part A, vol. *20*, pp. 252–256, June 1997.

x86 Weekly News, Via Reaffirms Support for DDR SRAM, x86 Weekly News, June 29, 2000, http://x86.ddj.com/news/2000/news062500.htm, accessed on January 18, 2002.

Xilinx, Additional Equivalent Wafer Foundries for Fabrication of Virtex(tm) Commercial (C-grade and I-grade) Product Family, PCN2000-08, Xilinx, December 29, 2000.

Zuehlke, K., Markt & Technik, Haengen wir am Tropf der Rohstofflieferanten?, http://www.elektroniknet.de/e-fertigung/news/article/29007/0/Hxxxaumlxxxngen_wir_am_Tropf_der_Rohstofflieferanten/, accessed on August 18, 2010.

ZVEI, Zentralverband Elektrotechnik- und Elektronikindustrie e. V. (ZVEI), Weissbuch "Langzeitversorgung der Automobilindustrie mit elektronischen Baugruppen," Frankfurt am Main, 2002.

ZVEI, Zentralverband Elektrotechnik- und Elektronikindustrie e. V. (ZVEI), RoHS—Restriction of the Use of Certain Hazardous Substances in Electrical and Electronic Equipment. Handlungshilfe zur Kommunikation entlang der Lieferkette ueber die Einhaltung stoffbezogener Anforderungen aus der Richtlinie 2002/95/EG (RoHS), 2008.

Index

Strategies to the Prediction, Mitigation and Management of Product Obsolescence, First Edition. Bjoern Bartels, Ulrich Ermel, Michael Pecht and Peter Sandborn.

WILEY SERIES IN SYSTEMS ENGINEERING AND MANAGEMENT

Andrew P. Sage, Editor

ANDREW P. SAGE and JAMES D. PALMER
Software Systems Engineering

WILLIAM B. ROUSE
Design for Success: A Human-Centered Approach to Designing Successful Products and Systems

LEONARD ADELMAN
Evaluating Decision Support and Expert System Technology

ANDREW P. SAGE
Decision Support Systems Engineering

YEFIM FASSER and DONALD BRETTNER
Process Improvement in the Electronics Industry, Second Edition

WILLIAM B. ROUSE
Strategies for Innovation

ANDREW P. SAGE
Systems Engineering

HORST TEMPELMEIER and HEINRICH KUHN
Flexible Manufacturing Systems: Decision Support for Design and Operation

WILLIAM B. ROUSE
Catalysts for Change: Concepts and Principles for Enabling Innovation

LIPING FANG, KEITH W. HIPEL, and D. MARC KILGOUR
Interactive Decision Making: The Graph Model for Conflict Resolution

DAVID A. SCHUM
Evidential Foundations of Probabilistic Reasoning

JENS RASMUSSEN, ANNELISE MARK PEJTERSEN, and
LEONARD P. GOODSTEIN
Cognitive Systems Engineering

ANDREW P. SAGE
Systems Management for Information Technology and Software Engineering

ALPHONSE CHAPANIS
Human Factors in Systems Engineering

YACOV Y. HAIMES
Risk Modeling, Assessment, and Management, Third Edition

DENNIS M. BUEDE
The Engineering Design of Systems: Models and Methods, Second Edition

ANDREW P. SAGE and JAMES E. ARMSTRONG, Jr.
Introduction to Systems Engineering

WILLIAM B. ROUSE
Essential Challenges of Strategic Management

YEFIM FASSER and DONALD BRETTNER
Management for Quality in High-Technology Enterprises

THOMAS B. SHERIDAN
Humans and Automation: System Design and Research Issues

ALEXANDER KOSSIAKOFF and WILLIAM N. SWEET
Systems Engineering Principles and Practice

HAROLD R. BOOHER
Handbook of Human Systems Integration

JEFFREY T. POLLOCK and RALPH HODGSON
Adaptive Information: Improving Business through Semantic Interoperability, Grid Computing, and Enterprise Integration

ALAN L. PORTER and SCOTT W. CUNNINGHAM
Tech Mining: Exploiting New Technologies for Competitive Advantage

REX BROWN
Rational Choice and Judgment: Decision Analysis for the Decider

WILLIAM B. ROUSE and KENNETH R. BOFF (editors)
Organizational Simulation

HOWARD EISNER
Managing Complex Systems: Thinking Outside the Box

STEVE BELL
Lean Enterprise Systems: Using IT for Continuous Improvement

J. JERRY KAUFMAN and ROY WOODHEAD
Stimulating Innovation in Products and Services: With Function Analysis and Mapping

WILLIAM B. ROUSE
Enterprise Transformation: Understanding and Enabling Fundamental Change

JOHN E. GIBSON, WILLIAM T. SCHERER, and WILLIAM F. GIBSON
How to Do Systems Analysis

WILLIAM F. CHRISTOPHER
Holistic Management: Managing What Matters for Company Success

WILLIAM B. ROUSE
People and Organizations: Explorations of Human-Centered Design

MO JAMSHIDI
System of Systems Engineering: Innovations for the Twenty-First Century

ANDREW P. SAGE and WILLIAM B. ROUSE
Handbook of Systems Engineering and Management, Second Edition

JOHN R. CLYMER
Simulation-Based Engineering of Complex Systems, Second Edition

KRAG BROTBY
Information Security Governance: A Practical Development and Implementation Approach

JULIAN TALBOT and MILES JAKEMAN
Security Risk Management Body of Knowledge

SCOTT JACKSON
Architecting Resilient Systems: Accident Avoidance and Survival and Recovery from Disruptions

JAMES A. GEORGE and JAMES A. RODGER
Smart Data: Enterprise Performance Optimization Strategy

YORAM KOREN
The Global Manufacturing Revolution: Product-Process-Business Integration and Reconfigurable Systems

AVNER ENGEL
Verification, Validation and Testing of Engineered Systems

WILLIAM B. ROUSE (editor)
The Economics of Human Systems Integration: Valuation of Investments in People's Training and Education, Safety and Health, and Work Productivity

ALEXANDER KOSSIAKOFF, WILLIAM N. SWEET, SAM SEYMOUR, and STEVEN M. BIEMER
Systems Engineering Principles and Practice, Second Edition

GREGORY S. PARNELL, PATRICK J. DRISCOLL, and DALE L. HENDERSON (editors)
Decision Making in Systems Engineering and Management, Second Edition

ANDREW P. SAGE and WILLIAM B. ROUSE
Economic Systems Analysis and Assessment: Intensive Systems, Organizations, and Enterprises

BOHDAN W. OPPENHEIM
Lean for Systems Engineering with Lean Enablers for Systems Engineering

LEV M. KLYATIS
Accelerated Reliability and Durability Testing Technology

BJOERN BARTELS, ULRICH ERMEL, MICHAEL PECHT, and PETER SANDBORN
Strategies to the Prediction, Mitigation and Management of Product Obsolescence

Printed and bound by CPI Group (UK) Ltd, Croydon, CR0 4YY
16/09/2026
14963423-0001